ANTICIPATORY WATER MANAGEMENT

USING ENSEMBLE WEATHER FORECASTS FOR CRITICAL EVENTS

T0186432

Anticipatory Water Management

Using ensemble weather forecasts for critical events

DISSERTATION

Submitted in fulfilment of the requirements of
the Board for Doctorates of Delft University of Technology
and of
the Academic Board of the UNESCO-IHE Institute for Water Education
for the Degree of DOCTOR
to be defended in public
on Tuesday, November 3, 2009 at 15:00 hours
in Delft, The Netherlands

by

Schalk Jan van ANDEL
born in Amsterdam, The Netherlands

Master of Science in Quantitative and Integrated Water Management with
Distinction, Wageningen University, The Netherlands

This dissertation has been approved by the supervisor
Prof. dr. R. K. Price

Members of the Awarding Committee:
Chairman   Rector Magnificus, TU Delft
Prof. dr. ir. A.E. Mynett   Vice-Chairman, UNESCO-IHE
Prof. dr. R.K. Price   Supervisor, UNESCO-IHE / TU Delft
Prof. drs. ir. J.K. Vrijling   TU Delft, The Netherlands
Prof. dr. ir. A.W. Heemink   TU Delft, The Netherlands
Prof. dr. ir. E. Schultz   UNESCO-IHE, The Netherlands
Dr. J.C. Schaake   NWS / NOAA, USA
Dr. ir. A.H. Lobbrecht   UNESCO-IHE, The Netherlands
Prof. ir. F.H.L.R. Clemens   TU Delft, The Netherlands, reserve member

CRC Press/Balkema is an imprint of the Taylor & Francis Group, an informa business

Published by:
CRC Press/Balkema
PO Box 447, 2300 AK Leiden, The Netherlands
e-mail: Pub.NL@taylorandfrancis.com
www.crcpress.com – www.taylorandfrancis.co.uk – www.balkema.nl
ISBN 978-0-415-57380-1 (Taylor & Francis Group)

Foreword

When in 2002, I learned of the existence of UNESCO-IHE, by that time still named 'IHE-Delft', I realised that this institute for water education encompassed all my professional interests and even some of my stronger personal interests (if this division between personal and professional interests really exists). UNESCO-IHE is entirely devoted to the aquatic environment, with emphasis on capacity building in less privileged countries. Perhaps its greatest merit is the 'peace building' through the cross national and cross continental friendships between around 200 new MSc students coming from all over the world every year. Walking through the corridors, and having lunch in the canteen, together with all these water professionals from such diverse background, is an honour and pleasure for me every day. I was convinced by that time, through a short period of introduction to different professional environments, that the combination of education and research is for me the most attractive way of spending one's professional life. Add to this my personal interest in water sports, such as windsurfing and wave-surfing, and it should be clear why UNESCO-IHE stands out as the most attractive working environment I can think of.

Two years later, in 2004, I learned from Dr. Lobbrecht, who is director of HydroLogic BV and also is a faculty member of the Hydroinformatics Core at UNESCO-IHE, what a challenge the highly variable weather conditions in the Netherlands are for the Water Boards, which are responsible for the daily operation of the regional water systems. Research was needed into the use of weather forecasts in decision support systems for operational water management. This research topic, involving Real-Time Control with its combination of water management and ICT, is one of the research areas of the Hydroinformatics and Knowledge Management Department at UNESCO-IHE. With the importance of flood and drought management, and the increasing availability and uptake of ICT in developing countries, the research was internationally relevant.

Acknowledgements

The research was funded by the Delft Cluster programme, the Principal Water-board of Rijnland and UNESCO-IHE.

The data for this research was kindly provided by the Royal Netherlands Meteorological Institute (KNMI), the Principal Water-board of Rijnland, National Meteorological Agency (NMA) of Ethiopia, Ministry of Water Resources Ethiopia, and Bahirdar metrological office. The HydroNET software for processing the meteorological data was kindly provided by HydroLogic BV.

I would like to thank Professor Roland Price for his enthusiasm for the topic, the ideas for exploration, the discussions on the content, the continuing efforts in correcting my English texts, the sharing of thoughts on philosophical, societal and theological issues, and for his personal way of supporting and guiding me. Professor Price's personal involvement with, care for, and support and guidance of students is an example for me in my further professional and private life.

With Arnold Lobbrecht I am working, not only on the current research, but also on many other project activities within UNESCO-IHE. Because of these always-ongoing busy activities, and because of the pleasant, informal way of working together, there has been little opportunity to express my appreciation and gratitude towards Arnold. Hence, I take the opportunity to do so here. I think Arnold is too modest about his vast knowledge and experience in water management and hydroinformatics.

This research would not have been possible without the countless inputs from professionals in the water management and meteorological fields. I would like to thank Frans van Kruiningen for his guidance throughout the research, for his insights into the Rijnland water system, and for his support in preparing journal papers. Also, my thanks go to thank René van der Zwan for continuing the cooperation between Rijnland and UNESCO-IHE after Frans joined the Principal Water-board of Delfland. Robert Mureau is the Ensemble expert at the Royal Netherlands Meteorological Institute (KNMI). My thanks go to Robert for the many discussions we had, for helping me to understand the ensemble prediction systems, and for his thorough reviews of journal papers. Robert Mureau is currently working with MeteoConsult. In addition, I appreciate the support of Kees Kok (KNMI) in understanding some of the verification techniques used in weather forecasting, and the continuing cooperation. Thank you also to Sander Loos and Timmy Knippers for their support with the HydroNET software, and to Jantine

Bokhorst for her support in processing the ensemble forecast data.

I have had the pleasure to work with MSc students on their thesis research work. In particular, I am grateful to Kibreab Amare Assefa with whom Chapter 5 about the Upper Blue Nile case study was prepared. His research work was funded by the WaterMill project.

I would like to thank Wilmer Barreto for providing me with the NSGAII software he developed. The software proved invaluable for my research.

Then, I would like to express my appreciation to Jan Luijendijk for introducing me to the Hydroinformatics and Knowledge Management Department, and for supporting me in finding the right balance between PhD work and the many other interesting activities available within the department. Thank you also to Professor Dimitri Solomatine for his support as the Head of the Hydroinformatics Core, for his input on the optimisation problems in this research, and for his friendly and humorous cooperation. I would like to thank my UNESCO-IHE colleagues Yasir Abas Mohamed, for introducing me to case study opportunities in the Upper Blue Nile, and Shreedhar Maskey, for his input concerning the decision-making challenges in flood early warning.

To Professor Michael Abbott, I am also very grateful for the many discussions on socio-economic and philosophical issues, and for introducing me to the Water Knowledge Initiative.

I would like to thank Andreja Jonoski for inspiring me in the work at UNESCO-IHE, on hydroinformatics and in general.

Thank you to Ioana Popescu, for the fine cooperation on other ongoing projects within the hydroinformatics group, and in particular for supporting me during the finalisation of this thesis.

I enjoy working together with my hydroinformatics colleagues: Professor Arthur Mynett, Professor Guy Alaerts, Zoran Vojinovic, Ann van Griensven, Biswa Bhattacharya, Yunqing Xuan, Carel Keuls, Judith Kaspersma, Giuliano Di Baldassarre, Jos Bult, and Gerda de Gijsel. In addition, I would like to thank my PhD colleagues and in particular Carlos Velez, Leonardo Alfonso, and Gerald Corzo, who have been working in associated fields of forecasting and control. Next to the scientific efforts, I have to recognise the not always successful sports team efforts with them. Thank you to all staff and participants of UNESCO-IHE for providing a great working experience.

Finally, I would like to thank the members of the doctoral examination committee for evaluating this thesis.

Summary

Most of today's inland surface-water systems are integrally connected to developments in human society. These systems depend on good day-to-day water management. Under normal operational conditions they present few problems. Critical conditions may cause problems, such as floods and droughts. These problems can be classified as having too much water, too little water, or water of poor quality. We try to minimise the frequency and extent of the damage due to critical events by water management. Strategic water management is concerned with catchment land use, spatial planning and water system design, while operational water management is concerned with the daily management of a given water system.

A large group of critical events are caused by meteorological extremes. Often operational water managers are informed too late about upcoming events to respond to them in an optimal manner. The lead-time provided by monitoring systems and hydrological predictions is not enough. Therefore, weather forecasting can be used, e.g. as input to the hydrological models, to expand the forecast horizon in water management. This is called Anticipatory Water Management (AWM). It allows water managers to take anticipatory actions to reduce the damage of critical events. An example of an anticipatory action is the lowering of a reservoir water level as part of flood control.

Similar to chess players when anticipating the moves of their opponents and planning their own counter moves, water managers can improve the performance of their systems, the more they are able to anticipate the upcoming events.

Hydro-meteorological forecasts are not always accurate. There is some degree of uncertainty whether the forecasted events will really occur. In particular, weather forecasts have a high degree of uncertainty, because the atmosphere is a chaotic system, in which small disturbances can grow rapidly to influence large-scale events. Anticipatory actions may, therefore, not be taken in time, or taken unnecessarily. Because of possible adverse effects of anticipatory actions, like the shortage of water for supply in the case of lowering of a reservoir water level for flood control, the uncertainty of the forecasts and associated risks of applying Anticipatory Water Management have to be assessed.

Ensemble Prediction Systems (EPS) have been developed to assess the dynamic uncertainty of weather forecasts. For each forecast the probability distribution is estimated by re-running the numerical prediction model with different initial conditions. This takes into account our limitations to measure or estimate the initial atmospheric state accurately at a high spatial and temporal resolution. The forecasted probability distribution allows water managers to make risk-based decisions. Much research focuses on providing reliable hydro-meteorological ensemble predictions. Increasingly water authorities and companies are making use of these predictions. This research focuses on the improvement of the end-use of ensemble prediction systems in Anticipatory Water Management.

A framework for developing Anticipatory Water Management strategies is proposed. *Firstly*, in this framework emphasis is given to the availability of hydroinformatics tools that allow flexible and realistic simulation of controlled water systems. Using these simulation models, the current water-management strategy can be emulated, and compared with alternative, anticipatory, strategies.

Secondly, it is emphasised that water authorities should themselves verify the performance of the hydro-meteorological forecasts local to their catchment. Generalised performance indicators, established on a regional or global scale as they are provided by meteorological institutes, do not provide sufficient information for local water management. The verification should be customised for the intended end-use of the Anticipatory Water Management. This means, for example, that the verification should focus on surface precipitation or rainfall-runoff modelling in applications to flood control. In addition, the verification should not be based on a fixed time interval, such as a day, but should establish for each event (e.g. an intense rainfall episode continuing for several days) whether or not it was predicted. The verification should be done using continuous time series and simulation, not only on the basis of a sample of critical events such as has been the practice in water management until recently. Only with continuous simulation can the full consequences of applying Anticipatory Water Management, including risks of false alarms (a forecast of a critical event while no critical event occurs) during normal conditions, be assessed. For this verification analysis, archives of water system data, meteorological data, and weather forecasts are needed. If an archive of weather forecasts is not available, than these weather forecasts need to be prepared. This can be done by re-running the numerical weather prediction models for the verification analysis period. This is called re-forecasting or hindcasting.

The water system model, together with the meteorological hindcast as input, allows "what-if" analyses for long periods. The analyses show water managers what would have happened if they had used the weather forecasts in their operational water management during the previous so many years. This already gives an indication of the effectiveness of the Anticipatory Water Management (AWM) in reducing the negative impact of critical events. For many water authorities, however, this will not be enough information to decide whether to adopt AWM. In most cases the (economic) efficiency should also be assessed. While general efficiency analyses are often performed using cost-loss ratios, these are not applicable to AWM because water management is highly dynamic. Each event is different and so are the cost-loss ratios.

Therefore, *thirdly*, a dynamic cost-model related to water system states, reflecting all the efficiency requirements, should be prepared by the water authority. Then the continuous simulation of water management can be translated into a time series of costs. The total damage costs of critical events and their development over the years can be assessed, and compared for different forecasting products and for different anticipatory management strategies.

If a forecasting product has been selected and the effectiveness and efficiency of AWM strategies are such that the water authority would like to adopt it, then as an extra step, an optimisation of the AWM strategy can be performed. The objectives of the optimisation in most cases would be to minimise the damage of critical events, and at the same time minimise the total damage. The standard risk-based approach, minimising the expected risk for every decision time step, may not be reliable, because it assumes the use of perfect probabilistic forecasts, while these are in reality not available. To take these uncertainties into account the strategy with the minimum costs over a long period (years) has to be found. This multi-year optimisation problem, in which per day several ensemble predictions are available and the best management strategy for the entire period needs to be defined, cannot be captured in an analytical optimisation model. Therefore, global optimisation methods with smart search methods, like evolutionary approaches, are used. Importantly, these search methods can be used with multiple objectives to provide a range of alternative strategies. This leaves the freedom to the water authority to select the optimal water management strategy depending on their perception of the importance of the different objectives.

The framework for developing Anticipatory Water Management strategies was applied to two case studies in flood early warning and control. One case study concerned a land-reclamation area in the Netherlands, Rijnland, and the other a tributary to Lake Tana in the catchment of Upper-Blue Nile, Ethiopia. The ensemble precipitation forecasts from the ECMWF Ensemble Prediction System and the NCEP Global Forecasting System (the frozen version for re-forecasting ensembles) were used. The ECMWF EPS is already received operationally by the Water Board in the Dutch case study. The NCEP GFS is freely available through the Internet and is therefore a very interesting research and operational tool for countries where investments for hydro-meteorological forecasting systems are not yet readily available.

For the Rijnland case study, effective warnings were obtained for most of the critical events in the analysis period. The optimisation of the Anticipatory Water Management strategy resulted in a 30% reduction of the estimated total costs, and a reduction of 35% of the flood damage costs over a 8-year period. This shows clearly that Anticipatory Water Management outperforms the traditional use of re-active operational water management. In the Netherlands, ECMWF EPS forecasts can be used to expand management horizons to three or more days. The fact that the optimal decision rules differ from the ones currently used by the Rijnland Water Board confirms the need for the Water Boards to perform hindcast analyses to improve their anticipatory management strategies.

The case study of the Blue Nile shows that freely available weather forecasts and hydrological modelling software can be used for research into prediction systems and Anticipatory Water Management strategies. For this particular case study, a warning could be obtained for a maximum of 60% of the peaks in the simulated reference streamflow (above flood threshold). The forecasting system needs further improvement before operational use is considered. For this improvement, bias correction and downscaling methods that are the focus of current international research efforts into Hydrological Ensemble Prediction Systems should be used. These methods to produce reliable probabilistic forecasts, with as small a predictive uncertainty as possible, will also be used in ongoing research to increase still further the efficiency of AWM for Water Boards in the Netherlands.

The backbone of developing successful and reliable AWM strategies is the verification analysis with continuous simulation spanning multiple years. Archives of weather forecasts of multiple years are necessary, because of the low frequency of critical events. These archives are generally not available because weather forecasting systems are continuously updated. Therefore, there is a strong need to prepare hindcast archives for new products. Because preparing these hindcasts interferes with the operational tasks of the meteorological institutes, the task of hindcasting should be relegated to separate, dedicated institutes. This would give a credible contribution to the practical use of weather forecast products.

This is important for water management worldwide, because it is clear that the performance of weather forecasts is such that water authorities cannot afford not to use this available information. This applies not only to the flood management case studies presented in this thesis, but for many more applications, such as drought management, and for many more types of water systems. Therefore, scientists and engineers are called on to join in an effort to expose and to cover the complete scope of Anticipatory Water Management, and to maximise the use of hydro-meteorological forecasts in operational water management.

Content

FOREWORD V

ACKNOWLEDGEMENTS VII

SUMMARY IX

1   INTRODUCTION 17

   1.1   BACKGROUND 17
       1.1.1   *Hydroinformatics and Integrated Water Resources Management* 17
       1.1.2   *Management of extreme events* 18
       1.1.3   *Operational water management* 19
       1.1.4   *Benefits of increased forecast horizon* 20
       1.1.5   *Use of weather forecasts* 21
       1.1.6   *Ensemble forecasts* 22
   1.2   ANTICIPATORY WATER MANAGEMENT 23
   1.3   HYPOTHESES AND OBJECTIVES 25
   1.4   READER 26

2   ANTICIPATORY WATER MANAGEMENT 27

   2.1   INTRODUCTION 27
   2.2   OPERATIONAL WATER MANAGEMENT 27
       2.2.1   *Definition* 27
       2.2.2   *Components of operational water management* 27
       2.2.3   *Water system control* 29
       2.2.4   *Reservoirs and polders* 31
       2.2.5   *Flood early warning and control* 32
       2.2.6   *Challenges in operational water management* 33
   2.3   WEATHER FORECASTING AND ENSEMBLE PREDICTIONS 34
       2.3.1   *Monitoring systems* 34
       2.3.2   *From hand-drawn weather maps to numerical prediction* 36
       2.3.3   *From deterministic to probabilistic forecasts* 37
       2.3.4   *Ensemble Prediction Systems* 38
       2.3.5   *Challenges in using weather forecasts for water management* 40
   2.4   MODELLING CONTROLLED WATER SYSTEMS 41
       2.4.1   *Definitions* 41
       2.4.2   *Model components* 42
       2.4.3   *Water system state prediction* 43
       2.4.4   *Challenges in modelling controlled water systems* 44
   2.5   DECISION MAKING WITH UNCERTAINTY 46
       2.5.1   *Uncertainty* 46
       2.5.2   *Risk* 47
       2.5.3   *Threshold-based decision rules for Ensemble Prediction Systems* ... 48
       2.5.4   *Cost-benefit analysis* 49
       2.5.5   *Decision Support Systems for Anticipatory Water Management* 50
   2.6   KNOWLEDGE GAPS AND HYPOTHESES 51

3   FRAMEWORK FOR DEVELOPING ANTICIPATORY WATER MANAGEMENT (AWM)..**55**

3.1   INTRODUCTION...55
3.2   ESTABLISHING THE NEED AND POTENTIAL FOR AWM.............55
  3.2.1   For which events is AWM needed.................................55
  3.2.2   Potential for anticipatory management action 60
3.3   VERIFICATION ANALYSIS..63
  3.3.1   Product selection: time scales, spatial scales.............63
  3.3.2   Continuous simulation of the real-time AWM forecasting system 63
  3.3.3   Event based verification of a range of decision rules for AWM 65
3.4   MODELLING CONTROLLED WATER SYSTEMS 67
  3.4.1   Input data based on end-use of model 68
  3.4.2   Framework for modelling controlled water systems.......68
3.5   STRATEGIES FOR ANTICIPATORY WATER MANAGEMENT 69
  3.5.1   Rule-based 70
  3.5.2   Pre-processing of ensemble forecasts to deterministic forecast.......71
  3.5.3   Risk-based..71
3.6   COST-BENEFIT OF SELECTED AWM STRATEGIES.......................73
  3.6.1   Dynamic cost-benefit analysis 73
  3.6.2   Sources of damage..74
  3.6.3   Anticipatory Water Management modelling.................74
3.7   OPTIMISATION OF ANTICIPATORY WATER MANAGEMENT.......75
  3.7.1   Objectives 76
  3.7.2   Parameterisation of AWM strategies...........................76
  3.7.3   Optimisation using perfect forecasts 77
  3.7.4   Optimisation with actual forecasts 77
3.8   DECISION MAKING FOR POLICY ADOPTION OF AWM..............78
  3.8.1   What-if analysis 78
  3.8.2   Re-analysis era..79
3.9   FRAMEWORK FOR DEVELOPING ANTICIPATORY WATER MANAGEMENT 79

4   CASE STUDY 1 - RIJNLAND WATER SYSTEM 81

4.1   INTRODUCTION...81
4.2   PROBLEM DESCRIPTION 83
4.3   DATA 84
4.4   WATER SYSTEM CONTROL MODEL...86
  4.4.1   Model structure...86
  4.4.2   Control strategy..88
  4.4.3   Model calibration 89
  4.4.4   Model validation...90
  4.4.5   Visualise what is not known and explain.....................92
  4.4.6   Modelling the unknown phenomena 95
  4.4.7   Final model results 97
  4.4.8   Discussion...102
4.5   ENSEMBLE FORECASTS VERIFICATION 103
  4.5.1   Precipitation ensemble forecasts archive..................103
  4.5.2   Water level hindcasts..103
  4.5.3   Event based verification for water managers 104
  4.5.4   Precipitation and water level thresholds 105
  4.5.5   Presently used precipitation threshold for anticipatory pumping ..105

     4.5.6   3-Day accumulated precipitation threshold for selected events *108*
     4.5.7   5-Day accumulated precipitation threshold for selected events *109*
     4.5.8   Discussion *111*
   4.6   ANTICIPATORY WATER MANAGEMENT STRATEGY DEVELOPMENT 114
   4.7   COST-BENEFIT OF SELECTED AWM STRATEGIES 116
     4.7.1   Water level - damage function *116*
     4.7.2   Inter-comparison of costs for selected strategies *118*
   4.8   OPTIMISATION OF ANTICIPATORY WATER MANAGEMENT STRATEGY . 120
     4.8.1   Optimisation with perfect forecasts *120*
     4.8.2   Optimisation with actual forecasts *122*
   4.9   ADOPTION OF AWM IN OPERATIONAL MANAGEMENT POLICY 125

5   CASE STUDY 2 - UPPER BLUE NILE **127**

   5.1   INTRODUCTION 127
   5.2   PROBLEM DESCRIPTION 127
   5.3   DATA 128
     5.3.1   Geographical data *128*
     5.3.2   Meteorological data *128*
     5.3.3   Streamflow data *130*
   5.4   HYDROLOGICAL MODEL 131
     5.4.1   Model set-up *131*
     5.4.2   Calibration and validation *132*
   5.5   ENSEMBLE FORECASTS VERIFICATION 135
     5.5.1   Event selection *135*
     5.5.2   Ensemble precipitation hindcasts *138*
     5.5.3   Ensemble streamflow hindcasts *138*
     5.5.4   Verification analysis *138*
     5.5.5   Statistical verification *139*
     5.5.6   Comparison by visual inspection *140*
     5.5.7   Flood early warning verification *142*
   5.6   ANTICIPATORY MANAGEMENT STRATEGY DEVELOPMENT 144
   5.7   ADOPTION OF AWM IN OPERATIONAL MANAGEMENT POLICY 145

6   CONCLUSIONS AND RECOMMENDATIONS **147**

   6.1   CONTRIBUTIONS TO ANTICIPATORY WATER MANAGEMENT 147
   6.2   DISCUSSION OF THE HYPOTHESES 149
   6.3   CONCLUSIONS 151
   6.4   RECOMMENDATIONS FOR MANAGEMENT PRACTICE 152
   6.5   RECOMMENDATIONS FOR FURTHER RESEARCH 153

REFERENCES **159**

LIST OF FIGURES **167**

ABOUT THE AUTHOR **173**

SAMENVATTING **177**

1   Introduction

1.1 Background

This research has been carried out under the auspices of UNESCO-IHE. UNESCO-IHE is a post-graduate educational and research institute entirely devoted to the aquatic environment. It is concerned with water resources management challenges worldwide, in particular, the challenges faced in less privileged countries. Every year, 200 water professionals from all over the word arrive in Delft, the Netherlands, to study for their MSc degree at UNESCO-IHE.

One of the most confronting new experiences these students report when arriving in the Netherlands, is the cold, rainy, and highly variable weather (the academic year at UNESCO-IHE starts in October). These same changing weather conditions also form a challenge for the Water Boards that are responsible for the daily operation of the regional water systems. At the start of this research in 2004, it had become apparent that although many water boards had installed, or were installing, Decision Support Systems (DSSs) with real-time weather monitoring and forecasting data and hydrological simulation models to anticipate better the changing weather conditions, many questions remained on how this wealth of information could then be used best in practice. How could Water Boards assess the quality of the weather information? How could they deal with errors in the data, and with uncertainties in the meteorological and hydrological forecasts? How should they make decisions to take anticipatory actions?

These same questions are just as relevant to any part of the world where extreme rainfall events or prolonged periods of limited rain may cause floods and droughts or agrevate water quality problems. The DSSs and hydro-meteorological forecasting tools are becoming readily available to developing countries. Therefore, the research fitted the mission of UNESCO-IHE in general and the objectives of the Hydroinformatics group at the institute in particular.

1.1.1 Hydroinformatics and Integrated Water Resources Management

Hydroinformatics (Abbott, 1991) is the science of information and communication technologies in integrated water resources management. Integrated water resources problems are complex, and ICT, including computer simulation models and computer presentation tools, helps water experts in their analysis. The use of this digitised, virtual world (Price, 2008) is often preferred over physical experiments in the real world, because risks associated with physical experiments in relation to water resources are too high, and time and budgets too limited. Measuring what is happening to the water resources is a prerequisite for informed management of these resources. Ongoing developments in real-time monitoring, both in ground station telemetry and remote sensing, and communication of the monitored data into fast and easily accessible data bases, have greatly enhanced the up-to-date information about the state of the water resources to be managed. Also, water experts are not the only people who need the help of ICT to analyse water resources. Just as important, and indeed still increasingly important, is the communication of information about the water resources and their management, to policy makers and the public. It is here that we realise that hydroinformatics is a socio-technology (Abbott, 1999). Developments in society, such as the full integration of the internet and mobile telephony worldwide, influence how ICT and Integrated Water Resources Management can best be combined. Hydroinformatics, in its turn, influences the way integrated water resources management is performed, increases the number of people involved and concerned, and as such influences society as a whole. Hydroinformatics, through the application of ICT, strives to make integrated water resources management available to even the least privileged societies.

1.1.2 Management of extreme events

Extreme events often (temporarily) unbalance the management of water resources. Many places on earth face extreme events, like floods and droughts, with devastating effects. In the Netherlands the most recent river floods occurred in 1993 and 1995, when 240,000 people and one million animals were evacuated. Economic losses amounted to more then 100 million US dollars (Moll et al., 1996; Boetzelaer and Schultz, 2005).

Whereas in the past, local problems due to extreme events may have been analysed in isolation, today, with the growing insights into the hydrological cycle and the interdependence of the different components of the natural and anthropogenic systems, the need for integrated water resources management becomes more important. The ability to analyse the water system at bigger spatial (catchments) and temporal (seasons, years) scales has improved considerably over recent years. It enables the water community to look for management practices that benefit both drought and flood management on the local and catchment scale. Part of these management practices is operational water management. This document presents a study of the mitigation of the negative impacts of extreme events by enhancing operational water management. Because there will always be events that are too extreme to be managed in any way, in this thesis we refer to critical events, to indicate the group of events that do permit mitigation actions.

1.1.3 Operational water management

First we have to understand the scope of operational water management. What would we do if we were responsible for today's operation of the Dutch Delta Works, or of the Dutch pumping stations? Is there a storm surge? Should we close the barriers? Will there be a rain storm tomorrow? Should we activate the pumps? This dissertation deals with such operational water management questions, and contributes to the enhancement of this management. It does not however, deal with the design of water systems and structural changes.

To stay focussed on operational water management, we first address some definitions; starting with the term itself.

> Operational water management is the set of day-by-day
> decisions and subsequent actions that interact with the
> water system.

In this dissertation, a system, according to Oxfords dictionary, refers to "a group of related things or parts working together".

> A water system thus becomes a set of water bodies with
> their conveyance and regulating structures that work
> together through natural and artificial processes.

Most of today's water systems affect people, and are affected by people. People are interdependent on water systems through their roles as beneficiaries, extractors of water, and, last but not least, managers. It should be noted that through the involvement of people, the whole will not behave in a systematic manner, and as a consequence the word "construct" is preferred over "system" (Abbott, 2005). However, because the use of the word "construct" is unfamiliar to most people, the word system is used in this dissertation. As water managers, people interact with the water bodies through the design and operation of structures, and with people themselves through, for example, water supply systems, consumption regulations, and, more incidental interactions, such as evacuation in times of flooding.

> The objective of operational water management is to
> maximise the benefits of water bodies for society.

According to developments in sustainable and integrated water resources management, it is now commonly accepted that the maximisation of benefits should be done in such a way that today's people, as well as future generations, benefit from the water resources. Through the objective of making optimal use of available water and water systems, operational water management inherently contributes to the benefits of water, taking acount both of today's climate, as well as a future, a changed climate.

Operational water management has gone through a long history of development. An overview of this development, with focus on the Netherlands, was given by Lobbrecht (1997, pp 4-13). In the area of the Netherlands active water management began around 800 AD with digging ditches to divert water. The developments afterwards, up to the 20th century, are mainly characterised by the expansion of available structures to manage water. First dikes, then dams, windmills, series of windmills, and finally, electrically powered structures became available. Examples in the Netherlands include controllable or regulating structures such as the Oosterschelde Storm Surge barrier and the Maeslandt Barrier to protect the coastal areas, and the hundreds of diesel and electric pumping stations to manage the land-reclamation areas (called "polders").

In the 20th century the technical advances in regulating structures levelled out, and the focus of technological research in water management moved towards the methodologies and means to operate all the structures efficiently. These developments profited strongly from developments in Hydroinformatics in general, and Real-Time Control (RTC) in particular.

The present state-of-the-art is the incorporation of developments in RTC with monitoring networks, communication systems, data bases, hydrological modelling software and decision support tools in complete hydroinformatic systems for maximising the benefits of operational water management.

Yet, with all these technological means in place, events still occur that cause much damage to the water system. It is the starting point of this dissertation that further reduction of these damages, within the constraints of the system's capacity, should be sought by enhancing the use of predictions of future states of the water system. It is by increasing the forecast and decision horizons, that water managers, like chess players who are able to think several turns ahead, can further increase the efficiency of their water systems.

1.1.4 Benefits of increased forecast horizon

Increasing the forecast horizon can be achieved, not only by using real-time monitored meteorological variables as input to hydrological simulations, but also by using forecasted meteorological variables as input to the models. Monitoring proved to be insufficient to provide the required forecast horizon to take management actions for part of the critical events. As a consequence, hydrological simulation models were developed and applied. These in turn do not always provide the required forecast horizon, hence the need for inclusion of forecasted meteorological inputs.

With the modelling studies, uncertainties in the predictions became apparent. The hydrological systems and certainly the meteorological systems that drive hydrological critical events are chaotic systems, in the sence that small changes in the present state lead to large changes in the future state of the system. Inherently, small errors in the initial conditions of a model lead to big deviations in the predicted states on the one hand and the actual states on the other. In response to this problem, ensemble modelling techniques have been developed, and their further enhancement and use in practice is today at the forefront of scientific research (Schaake et al., 2006).

1.1.5 Use of weather forecasts

Developments in numerical atmosphere modelling and atmosphere remote sensing have resulted in a readily available suite of meteorological products for water professionals. National weather services offer model output time series and images directly, e.g. through File Transfer Protocol (FTP), to water management agencies that can automatically process and forward these time series as input to hydrological models for decision support. Next to this increase in real-time availability of meteorological data, another important development is the use of re-analysis and hindcasting. Hindcasting means that when a new meteorological product becomes available a data set is prepared of what would have been the results of the product if it had been used for the past so many years. This data set can be used to compare a new product with the old products and to train the use of the new product.

The need or significance to include weather forecasts in the preparation of water system predictions differs per application and type of water system (Figure 1.1). For management actions that need little time to become effective, like the control of weirs and gates in irrigation canals, not much lead-time is required and therefore the significance of weather forecasts is less. Early warning and evacuation measures take a long time to become effective, but for large rivers long forecast horizons can be achieved using upstream measurements and river simulation models, without using weather forecasts. Therefore, also in this case the significance of weather forecasts is limited (Figure 1.1). For flood control measures that need a long time to become effective, like lowering reservoir levels, in fast responding catchments and in catchments where flooding problems follow directly from extreme rainfall events (pluvial flooding), the significance of the use of weather forecasts is very large. Also for drought management, where seasonal forecasts are needed, the significance of long-range weather forecasts (rainfall and temperature) is large. In Figure 1.1 a number of application areas for the use of weather forecasts in operational water management have been tentatively positioned according to their required lead-time and the significance of the meteorological forecast.

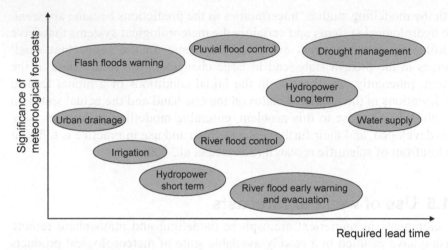

Figure 1.1 Significance of meteorological forecasts for operational water management applications

1.1.6 Ensemble forecasts

Ensemble forecasts are forecasts that contain a number of alternative predictions for the same forecast period. One such prediction is called an ensemble member. The differences between individual members can be the result of differences in expert opinions, in atmosphere simulation models used, or in the initial conditions used for the models, depending of the kind of ensemble system that was used. In any case, the differences in the ensemble members provide information about the uncertainty of the particular forecast. If all the members are more or less the same, the forecast has a measure of certainty about what is going to happen. If, on the other hand, the members show large differences, it means that the forecast is highly uncertain.

Ensemble forecasting to provide this real-time estimates of forecast uncertainty has become common practice for the bigger international meteorological organisations, such as the European Centre for Medium range Weather Forecasting (ECMWF) and the National Centre of Environmental Predictions (NCEP). The ensemble forecasts have become available to more and more countries at lower costs or free of charge.

These ensemble weather forecasts can be used to expand the decision horizon of operational water management to anticipate what may happen.

1.2 Anticipatory Water Management

Anticipatory Water Management (AWM) is defined as daily operational water management that pro-actively takes into account expected future conditions and events. "Future events" refers to events that are not yet measurable within the catchment. Therefore, weather forecasts must be applied to prepare the predictions. Examples of Anticipatory Water Management actions are the lowering of reservoir levels for flood control and maintaining reservoir levels in anticipation of droughts.

We apply anticipatory management ourselves in every day life. If we expect the train to be delayed, we take an earlier train to be on time. If we expect the weather to be warm we dress accordingly. If Johan Cruijf expects the Dutch soccer team to win, we put our money on the team. In each example we make an estimate about the credibility, assess the risks and make a decision. Johan Cruijf is regarded as the expert on soccer and thus we put a lot of faith in his forecasts and take the risk of losing money.

Also in professional fields, predictions of all kinds are used in management. Economic and market forecasts are used in organisation management and product development. Demographic development models and climate change predictions are used in land use planning, and meteorological forecasts are used for natural hazard warnings in agriculture, transport (aviation, shipping, road traffic), defence and healthcare (NHS, 2002).

Applications of meteorological forecasts mostly concern short-term forecast horizons. If tonight's temperatures are forecasted to be below zero, farmers protect their crops against freezing. Storm warnings for shipping and aviation only apply to the present day, and action in many cases is taken only when the forecasted event is already taking place.

The reason that mid-term and even short-term weather forecasts are in many cases not decisive in daily management is that these forecasts are considered not to be accurate enough. As a consequence their weight in the decision making process is often very small (or the forecasts are not used at all), and the role that forecasts should play in the decision process is not formalised. It is left to the judgement of experts and managers what to do with them. In the case of the farmer, wetting his crops to protect them from freezing, this is not a problem, because the costs of the action are low. If the weather forecast turns out to be wrong, the economic damage to the farmer is little. If the crop is lost due to frost, the damage is huge, so the farmer's risk analysis leads to a clear decision to protect his crops.

With daily water management, especially when concerning critical events, a risk analysis of the use of an uncertain forecast is more complicated. Costs of pro-active management actions like an evacuation and controlled flooding are high and thus a decision is delayed as long as possible. Lead-times of management decisions and response times of actions (like an evacuation) play a crucial role. In order for pro-active management to be effective the decision has to be made before the actual critical event is due within the response time of the management action.

Despite these difficulties, there are currently many reasons to put effort into improving Anticipatory Water Management:
- Global annual loss of life and socio-economic costs due to extreme events are still very high, and even increasing;
- Next to long term (structural) prevention and mitigation strategies, there is a need to do what we can with the water systems that we have now;
- Ever increasing human pressure on natural resources and growing economic constraints call for optimal use of available water systems, before large scale and expensive structural changes are made (WB21, 2000, p. 51);
- New meteorological observations and forecasts have been (further) developed, such as radar, satellite and, very importantly, ensemble forecasts;
- Qualitative and quantitative water-system response models have improved. Recent hydrological science has put effort in uncertainty analyses of these models, enabling risk analyses and, therefore, better informed decision making;
- Anticipatory management actions for critical events are available, like lowering storage reservoir levels for flood control and maintaining water levels to prevent droughts;
- In many different fields scientific and practical progress has been made on risk analyses and decision-making.

The basic process of Anticipatory Water Management distinguishes four steps. First, the present state of the water system has to be determined, and at the same time meteorological forecasts of atmospheric variables, such as precipitation and temperature, have to be acquired. Then the response of the water system to the atmospheric variables can be predicted, resulting in a forecast of the state of the water system. On the basis of all the information acquired, operational management decisions have to be taken, taking into account the uncertainty of the forecasts and the risk of each management option. If management actions are required, these need to be implemented. This basic concept of Anticipatory Water Management is illustrated in Figure 1.2.

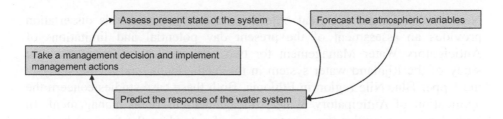

Figure 1.2 Basic representation of the process of Anticipatory Water Management

1.3 Hypotheses and objectives

The main incentive and objective of this dissertation is to:

> Improve the use of weather forecasts in operational
> water management.

The present ways of using weather forecasts in operational water management are analysed, challenges in using weather forecasts identified and methods to meet these challenges are presented. These methods will contribute to more effective pro-active operational measures, such that operational practice can add to its real-time function a focus on Anticipatory Water Management.

The main hypotheses that are proposed in this dissertation to support the main objective are:

> The use of ensemble precipitation forecasts to decide on
> anticipatory control actions, in preference to re-active
> control, can reduce the damage costs over a long period
> of time.

> Long-term simulation of the complete Anticipatory
> Water Management strategy for a historic time series
> enables an optimisation of the strategy.

Firstly, these hypotheses reveal a focus on recent developments in ensemble forecasting in operational meteorology. Verification methods from meteorological sciences have been applied and adjusted to analyse these ensemble weather forecasts and water level forecasts derived from the ensembles. Secondly, the focus of the methodologies is on the capitalisation of enhancements in hydroinformatic systems, such that extensive computer simulation of the operational water management strategies can be performed to develop and evaluate novel strategies.

Next to these methodological objectives and hypotheses, this dissertation provides an assessment of the present day potential and limitations of Anticipatory Water Management for two case studies: an extensive case study of the Rijnland water system in the Netherlands, and a case study in the Upper Blue Nile region in Ethiopia. Both these case studies concern the application of Anticipatory Water Management in flood management. In both these case studies the consequences of extending the forecast horizon from a maximum of 1-day, to 3-days or more are analysed.

1.4 Reader

In Chapter 2 a literature review is presented to identify knowledge gaps and hypotheses (listed in Section 2.6). In Chapter 3 theoretical and methodological concepts are explored to develop a framework for enhancing Anticipatory Water Management. In Chapter 4 and 5 this framework is applied to the two case studies. The dissertation concludes with a discussion on the hypotheses, and conclusions and recommendations for application of and research on Anticipatory Water Management.

2 Anticipatory Water Management

2.1 Introduction

In this chapter the different aspects of Anticipatory Water Management are reviewed. The introduction to operational water management is elaborated further. Weather forecasting, water system modelling, and decision making under uncertainty, are discussed in detail as key issues in Anticipatory Water Management.

2.2 Operational water management

2.2.1 Definition

Operational water management has been defined in chapter 1 as the set of day-by-day decisions and subsequent actions that interact with the water system. Operational water management concerns the daily control of water systems. It is performed to try and prevent water from threatening human life and to optimise its use for functions we consider important. Operational management is not concerned with the development of policy guidelines and the structural design of water systems. In the first place, monitoring and issuing early warnings in the event of calamity threats are very important tasks. Secondly, in most modern systems, management involves operating several regulating structures to minimise the frequency of calamities while the requirements of stakeholders are adequately (or optimally) met.

2.2.2 Components of operational water management

Operational water management consists of many components and tasks. The basic elements are:
- Structure and facilities
- Monitoring
- Objectives
- Management

The structure and facilities are, e.g. the river beds, canals, and embankments that contain or convey the water body, and the regulating structures, such as pumping stations and weirs. The monitoring includes everything that (contributes to) the provision of up-to-date information on the state of the water system. The word 'objectives' refers to the requirements of people, the ecology, and the water body itself. Present day operational water management has to take into account the requirements for these other beneficiaries (ecology and water systems as such) as well. Even remote water systems, like rivers where no people are living, are nowadays partly managed to preserve the right conditions for the ecosystems which the rivers are part of. Without requirements that are not continuously met by self-regulation of the water body, there is no need and no direction for operational water management. When concerning critical events, requirements are, for example, the maximum frequency of damage occurring, or the acceptable total damage over a certain period.

Management is done by the people responsible for making the daily set of decisions and the actions concerning the operation of the water system. Note that the decision makers and those taking subsequent action are grouped here, because theoretically it could be just one person doing all the work. In reality in almost all water systems the tasks of managing and implementation are divided between several people, and implementation is often done through automatically controlled structures. In the case of flood warning and evacuation, for example, there is usually a team responsible for the decision of issuing the evacuation order, while other people are responsible for the communication and execution of this order. In the end, even the evacuees themselves play an important role in making the evacuation effective. On the other end, there is the example of an automatic weir. Such a weir is controlled automatically (control actions are received from automatic functions, called controllers) and implements the regulating action, every 5 minutes, say, without interference of a manager. The task of the operational manager here is to check whether the settings of this automatic process are still effective, or need to be adjusted.

Next to elements of operational water management, tasks of operational water management can also be defined. These are:
- Development of a management strategy
- Daily management, operation and control
- Maintenance
- Evaluation of the operational management

With daily management we refer to the operational decisions that have to be taken on a regular basis. The frequency is at least daily for almost all systems, but often can be hourly or even every minute. So the term "daily" does not exclude other frequencies here, but is used because it clearly refers to the regularity of the decisions to be taken, as opposed to the determination of the management strategy.

The present research focuses on the management strategy and the daily management. It does not focus on maintenance, because for the case studies the existing water systems, including their design, regulating structures, monitoring networks and requirements are considered as given. Within the daily management the focus is on the information and decision support systems that can be used to optimise the decision making. The process of actually taking the necessary actions after the decision is made is also considered, to ensure that the proposed strategies are realistic.

In operational management there are many differences between management of a mainly natural system, such as flood warning and evacuation for large rivers, and management of strongly controlled water systems, like an urban water distribution network. The operational management of the latter is referred to as Water System Control. The major difference is the presence of regulating structures.

2.2.3 Water system control

In Chapter 1 the shift in development of water system control in the Netherlands from developments of control structures (starting from 1200 AD) to the development of control strategies (in the 20th century) has already been described. Most research and development has been done on these control strategies and is still being performed within the research community of Real-Time Control.

In the present day, the term "Real Time Control" is widely used. At first sight it seems like a self-evident term for something that has already existed for a long time. Operational water managers always make their decisions in real time. However, the "time" in "Real Time Control" does not reflect the absolute time of action (past, present or future), but the time origin of the data used and the moment the decision is made relative to the event of concern. Schilling (1990) defined real time control as *"a synonym for the manipulation of a process during its evolution"*. Traditionally control strategies are pre-set following historic system analyses. Sophisticated time series analyses and state-of-the-art, physically based scenario models will continue to be used to define the boundaries and guidelines for water system control. Examples are heights of river dikes, upper and lower boundary reservoir levels, and water allocation schemes. In contrast, real time control uses system state, prediction and user demand data that are as up-to-date as possible, and then seeks the optimal control strategy. Inherently the control strategy is continuously updated. Real time control seeks the best control strategy for the given moment within the constraints set by historically based boundaries.

The desirability to control in real time, or better "as soon as possible", has resulted in developments in several areas. Measuring devices have been modernised. Telemetric networks, radar and satellites have dramatically increased the data availability and acquisition speed. Data assimilation has become a very important research area. It focuses on state-of-the-art data processing techniques, optimising their use in computational models.

Despite the development of the computational power of computers, the time taken to solve the optimisation problem for RTC using these computational water system models can become the limiting factor. In response much research has successfully adopted machine-learning techniques to replicate hydrological modelling components using an order of magnitude less computational time (Lobbrecht et al., 2002; Bhattacharya et al., 2003; Lobbrecht and Solomatine, 1999).

Feedforward-feedbackward control
Real-time control has moved from simple manual control to complex multi-objective automatic control. When the system is controlled based on measurements of the target variable it is called feed-backward control. The advantage is that the effect of the control action (e.g. pumping) is monitored. When the system is controlled based on measurements of disturbances (e.g. rainfall) and the modelled effect on the target variable, it is called feed-forward control. Because the target value itself is not measured, combined feed backward-feed forward control is usually preferred in water management.

Model Predictive Control
Originally, in feedforward-feedbackward control the control for each time step is aimed at minimising the deviation from the target variable value by counter balancing the predicted effect of a measured disturbance. In a drainage system, for example, the predicted run-off for the coming hour is equalled by a control response of a pumping station by a discharge of the same amount, to keep the water level unchanged. This works fine for many cases and is still used in many controlled water systems. However, in cases when the control capacity is not enough to compensate the effect of the disturbance, the target variable value cannot be maintained. Then an intended deviation from the target variable value in opposite direction may be desired to limit the maximum deviation over several control time steps. To perform this automatically a control algorithm (controller) is needed that takes into account constraints of the water system, and is able to perform an optimisation on the basis of objective functions to assess whether, when and to what extent intended deviation from the target variable value is desirable. A widely known controller that has been developed to accommodate these requirements is Model Predictive Control (Overloop, 2006; Weijs et al., 2007).

Global control
In many cases, when facing extreme hydrological loads, not all of the water systems storage capacity is used at the moment of failure. Especially in sewer-system engineering, global (or central) control systems have been used and have shown that flood management can be improved by re-allocating water to other sub-systems before the most critical moment. To achieve this, in global control, several structures, controlling different

sections of the water system, are operated in such a way that the water is optimally distributed within and discharged from the system. This is an enhancement compared to local control, where a structure is operated based only on the system state in its direct vicinity.

Dynamic control
The adjustment of the control of the water system with changing spatial and temporal requirements is called dynamic water management or dynamic water system control (Lobbrecht, 1997). An example is taking into account the seasonal change in the requirements and risks of the agricultural sector.

Regardless the type of control, effectiveness depends on the available storage or the throughflow capacity in the water system. Reservoirs and polder are examples of systems that typically have a lot of storage.

2.2.4 Reservoirs and polders

In rivers, irrigation systems, and drainage systems the regulation often involves reservoir control. Large river reservoirs traditionally have been subject of intensive optimal control studies and practices. This is mainly due to the high economic value of the reservoirs. The physical properties of the reservoir are well defined such as the geographical boundaries, water volume and water level. Control is usually straight forward using sluices to adjust the reservoir water level. Determining the optimal real-time operation, however, is far from easy. Most reservoirs have multiple functions such as hydropower generation, water supply (irrigation, drinking) and safety against flooding (multipurpose reservoirs). Therefore multi-criteria or multi-objective functions have to be satisfied to arrive at the best operation. The most difficult problem, however, is the prediction of the future state of the reservoir. This involves meteorological and hydrological calculations throughout the entire upstream catchment and the determination of user demand.

Due to developments towards integrated water management in which the catchment hydrology plays a central role, research has expanded to include multi-reservoir operations (Huang et Yuan, 2004). The focus of these authors is on drought early warning and the practical use of real-time control.

In irrigation and drainage, the system of canals together forms a special kind of reservoir. In the Netherlands these regional irrigation and drainage systems have received their own terminology. "Polder" is a Dutch word referring to a low-lying land area that has been reclaimed from the water (lakes or seas). Water is drained from the land and pumped into the main conveyance channel, which is called a "Boezem". The excess water from this reservoir is discharged by pumps, weirs or sluices to rivers or the sea. Schultz, 1992, provides an historic overview of design and water management of the Dutch polders.

Polder management has made a shift, and is still in transition, towards integrated water management. Water systems are considered in their entirety and interests of different stakeholders are being taken into account (Bhattacharya et al., 2003). There is also more emphasis on seeking economic and optimal tailor-made solutions (Schultz, 1992; Wandee, 2005), whereas previously, average-based, policy guideline solutions were sufficient. This results in multi-objective or multi-criteria approaches in which various interests are balanced while minimizing damage costs (Lobbrecht et al., 2002).

In reservoir and polder management flood early warning and control often plays an important role.

2.2.5 Flood early warning and control

The present research has chosen applications in flood management because of the high relevance for the Netherlands and many developing countries, and because of the strong developments in Quantitative Precipitation Forecasting for the short to medium range (up to 15 days).

Therefore the focus is on rainfall induced flooding in fresh water systems. The flooding occurs because the hydrological load exceeds the capacity of the water system. The capacity is a combination of the discharge capacity and the storage capacity. Depending on whether and which structures are present, the discharge capacity consists of river or channel flow capacity and discharge structure capacity (e.g. pump or sluice), and the storage capacity consists of the "in bank" storage and the (emergency) storage basins and reservoirs.

Effective flood management integrates structural and non-structural measures, and long-term planning and operational preparedness (Price, 2006). Possible measures to reduce the negative consequences of flooding are to move human and economic activities out of the flood prone areas permanently or through early warning and evacuation. Non-structural measures to reduce the frequency and magnitude of this kind of flooding, consist of changes in land use that increase the water retention capacity of upstream parts in the catchment, for example, by planting forests to prevent erosion and subsequent surface runoff. Structural measures to reduce the frequency of this kind of flooding are to increase the discharge capacity of the control structures or the storage of the water system.

All these measures take a long time before they become effective, and are very expensive and spatially demanding. In today's world generally time, money and space are scarce. Hence it is important to also optimise water management within the current water system capacity. Maximising the use of the current water-system capacity is even more important when flooding problems are urgent or when structural measures are simply not sufficient to meet safety standards.

To further improve flood early warning and control, decisions have to be made on the basis of weather forecasts. This is defined as Anticipatory Water Management (AWM). Where the application concerns the operation of regulating structures it is called anticipatory water system control, or just anticipatory control.

2.2.6 Challenges in operational water management

Most of today's operational water management is still re-active or includes only hydrological predictions in feed-forward or model predictive types of management. Yet, at the same time, many water authorities and enterprises already receive and use weather forecasts and their number is increasing rapidly. While this expert based experience is already there, publications on these experiences are limited. Especially questions remain about how to deal with the high level of uncertainty when dealing with weather forecasts (Lobbrecht, 1997; Overloop, 2006). Therefore, there is a need on research to the end-use of weather forecasts in operational water management, which can result in a framework to develop and evaluate Anticipatory Water Management strategies.

This research is application oriented. Application oriented research involves not only academia, but also practitioners. The first group relies on theorems and scientific research, while the latter group is used to adopting a systems' approach in which confidence in a new system is built up over time through experience and feed back loops. In water resources engineering, and in this research in particular, such confidence is translated in the trust hydroinformaticians put in the modelling of processes, while operational managers trust in running extensive system tests. Practically oriented research should accommodate both requirements. Therefore Section 2.2 discussed the different operational water management practices, while Section 2.3, weather forecasting, and Section 2.4, water system modelling, address the processes and models that govern the systems' operation. In Section 2.5, decision making, we discuss how the academic and the system approach can be brought together to foster change. In particular re-analysis and verification as empowered by modelling systems are elaborated throughout the remainder of the dissertation as the vehicles for bridging the gap between theory and practice.

2.3 *Weather forecasting and ensemble predictions*

Currently the traditionally distinct disciplines of meteorology and hydrology are moving towards each other. Meteorologists try to incorporate the needs of hydrologists in their research programmes and hydrologists look for ways to optimise the potential of meteorological data (Lobbrecht and Loos, 2004; Lobbrecht et al., 2003; WMO, 2004).

Synoptic meteorology and climatology deal with the weather on small and large (averaged) time scales respectively. For operational water management the actual and the expected near future (maximum one year ahead, if we do not consider planning of groundwater extractions and such like) states of the water system are most important. Therefore the focus of this section is on synoptic meteorology, e.g. weather measurements and weather forecasts.

2.3.1 Monitoring systems

The backbone of meteorological science is still the global monitoring network of ground stations, ships, data buoys and radio sondes, (almost) directly measuring state variables such as temperature and wind. Many other variables and measurement techniques require translation of the measured parameter into the state variable, e.g. from satellite derived cloud-top temperature to precipitation. Together with the interpolation from point or local measurements to a 4-dimensional description of the state of the atmosphere, this translation is called weather analysis.

In addition to the classical measuring devices such as thermometers for temperature and tipping buckets for precipitation, the last decades have seen the development of modern techniques like automated weather stations, radar and remote sensing from satellites. The great advantage of the latter two over local measurements is their spatial coverage. Satellites can have global coverage, thus filling significant gaps in the observational network.

Weather satellites
The first satellites were launched in the early 1960's to make visual and infrared images of the earth. Polar satellites orbit Earth in a north-to-south direction at relatively low altitudes and obtain images of the entire globe in 12 hours. Geo-stationary satellites remain at a fixed point above Earth at the expense of a greater distance and therefore of detail on the images. Most applications still involve visual tracking of weather systems and potential rain storms, but intensive research is going on to enhance the quantitative analyses of satellite information. Wind speed can be estimated from cloud movement and temperature at different heights as deduced from radiation measurements of specific wavelengths.

Weather radar and now-casting
The use of RAdio Detection And Ranging (RADAR) for weather analyses has increased considerably. Conventional radar sends out electromagnetic waves and detects the reflected proportion and delay. These can be interpreted as a measure of the rainfall intensity and the distance of the storm respectively. Doppler radars also measure differences in frequency of the sent and reflected pulse. An increase of frequency indicates movement of the storm towards the radar. Two or more Doppler radars can therefore determine the direction and speed of storms. This technique is used for tornado tracking in the United States. When applied near real time it is often called now-casting (Lutgens and Tarbuck, 2001, p. 293, p. 329).

Rain radar provides users with the spatial variability of precipitation on a resolution that is almost never met by ground station networks. These ground stations can be used for calibration. The Royal Netherlands Meteorological Institute (KNMI) operates two Doppler radars and provides calibrated radar precipitation sums every 24 and 3 hours and non-calibrated sums every 5 minutes, both on a 2.5 km and a 1 km grid.

The most recent technical development is the positioning of a weather radar on board of a satellite. The Tropical Rainfall Measurement Mission (TRMM) is the most famous example. Japan and the United States work together on this project. The Satellite covers the tropical band (35N to 35S) and the project provides several rainfall analysis products, such as real-time 3-hourly precipitation estimates (NASA, 2008). Another feature of this project marks a very important development in meteorology, namely that the products and research results from this satellite based weather radar are freely accessible through the Internet. The increasing availability of meteorological data at no or low cost is a land mark development, that benefits science and society in general, and developing countries in particular (Akhtar et al., 2009).

Data availability and assimilation
With the growing measuring network and techniques, the available data has increased tremendously. Fortunately techniques of gathering, storing, processing and presenting this data have also been improved. Automated weather stations communicate their data to a central computer based system (telemetry), without intervention of human observers (Lobbrecht and Loos, 2004, Lobbrecht et al., 2003). Data assimilation techniques are used to model accurate initial fields for numerical weather prediction (Falkovich et al., 2000).

The present research uses rain radar and satellite derived radar indirectly through the use of numerical prediction models, but also directly in the hydrological modelling for the case studies. Ground based rain radar from the Royal Netherlands Meteorological Institute (KNMI) is used for the

Rijnland case study (section 4.3) and TRMM data is used for the Upper Blue Nile case study (section 5.3.2). For both case studies ground station measurements are used as well.

2.3.2 From hand-drawn weather maps to numerical prediction

A prediction of the state of the atmosphere at a certain time and certain place can be based on observations and knowledge of past events and physical relationships. This is called a weather forecast. The state variables that are of most concern to water managers are wind direction and speed (for open water levels), evaporation (water balance) and most of all, precipitation. Therefore Quantitative Precipitation Forecasting (QPF) is a lively scientific discipline on its own. Precipitation typically results from either small-scale convective weather systems or large-scale systems (fronts). The first is more difficult to forecast, because it concerns local events of high intensity precipitation.

Mostly three major groups of forecasts are distinguished based on their lead-time. The classification is not universal but in many cases up to 2-days ahead forecasts are referred to as short-term forecasts. Medium-range forecasts concern lead-times up to 10 days ahead, and long-range forecasts concern monthly and seasonal forecasts (Persson and Grazzini, 2007).

Meteorological science has seen the development of forecasting methodologies from synoptic forecasts, via deterministic numerical weather prediction (NWP) to probabilistic forecasts.

Synoptic forecasting

Traditionally the collected data are presented on maps. These synoptic weather maps, containing amongst other features isobars and wind direction and speed, are then used by meteorologists to extrapolate future conditions. Over the years, as experience had grown, the analogue forecasting method became the most important. Forecasts were made by comparing current patterns with patterns from the past. From this experience, rules of thumb were developed that still serve an important role in short term forecasting. The expertise of individual forecasters is of decisive importance for the quality of synoptic forecasts.

Numerical weather prediction

In numerical weather prediction processes and dynamics of the atmosphere are mimicked using physical laws. The gas law and the hydrostatic law describe the static relation between variables (diagnostic), and the equations of continuity, motion and the first law of thermodynamics describe the dynamic changes (prognostic). More specific processes are described using parameterisation schemes, such as a cloud scheme. The idea of numerical weather prediction had already been posed in 1904 by Vilhelm Bjerknes, but only in the 1960's were the theory sufficiently developed and the computational power (computers) advanced enough to come up with the first general circulation models. To solve the equations the atmosphere is discretised in vertical layers and a horizontal grid. Together with the required time step, suitable numerical schemes have to be chosen to process the forecast (Persson and Grazzini, 2007; Lutgens and Tarbuck, 2001).

Based on the spatial coverage of the models a distinction could be made between Global (Circulation) Models and Limited Area Models or Local models. Operational global models are hosted, for example, by the US National Centres for Environmental Prediction (NCEP) and the European Centre for Medium-Range Weather Forecasts (ECMWF). The ECMWF model has a spatial resolution of approximately 25 km and a time step of 15 minutes. The model output time step is 6 hours and the forecast horizon, or lead-time, is 15 days.

National weather services develop their Limited Area Models (LAM). The boundary conditions for these models are mostly provided by the global models. The LAM is then nested in the global model. The Royal Netherlands Meteorological Institute uses the HIRLAM model, which has a spatial resolution of 11 km, an output time step of 1 hour and a lead-time of 1 day. For certain applications, such as flood forecasting in mountainous areas, these resolutions are still not sufficient. Therefore downscaling techniques are being developed to try and bridge this gap (Ferraris et al., 2003).

2.3.3 From deterministic to probabilistic forecasts

The models described in Section 2.3.2 produce a single valued output for atmospheric variables (i.e. precipitation depth) at a certain place or area and at a certain moment in the future. This is called deterministic forecasting. In reality these forecasts are uncertain and therefore will often contain errors. Because of these errors, and perhaps to a greater extent because of the lack of knowledge and communication of the uncertainty in the predictions, the general public and end-users put little trust in the forecasts. Moreover, for proper decision-making, the uncertainty has to be known to carry out a risk analysis.

Already in 1965 precipitation probability forecasts were issued. Probability refers to the chance that an event will occur and is represented as a number between 0 and 1 or as a percentage (Lutgens and Tarbuck, 2001). The probability is mostly estimated by expert forecasters, based on the above described synoptic maps, deterministic model output and tracking tools like radar and satellite images (Atger, 2001). Interpretation can be very difficult and presentation by the forecasting bureaus misleading. In the Netherlands for instance a common way of presenting precipitation forecasts is by giving a precipitation depth in mm and a probability. Many people do not know that the precipitation depth is the deterministic output of HIRLAM and the probability refers to the chance that any amount of precipitation will fall at any point in the area during the forecast time period. The probability is therefore not related to the precipitation depth.

To overcome these difficulties methods have been developed that are all based on a comparison of several model outputs for the same forecast. If the model outputs are similar it is expected that the probability that the forecasted event will occur is high. If the model outputs are very different from each other the probability is low. Apparently the particular weather system is difficult to predict by the models. Examples are the use of different LAMs and running the same model using different initial conditions or parameterisations (Lobbrecht et al., 2003; Atger, 2001). The latter is called the Ensemble Prediction System (EPS) and accounts for the chaotic behaviour of the atmosphere ("butterfly effect"). Chaotic phenomena occur in deterministic systems that are very sensitive to initial conditions. Because of the chaotic behaviour, small errors in the determined initial state of the system will result in large errors in the forecasts of the future states of that system.

2.3.4 Ensemble Prediction Systems

Because of the sensitivity of the atmosphere to initial conditions, the present state of the atmosphere needs to be known up to a high spatial resolution to make accurate forecasts. Because it is impossible to monitor the whole atmosphere with required accuracy and spatial resolution, weather forecasts are often faced with much uncertainty.

The initial conditions are assessed as accurately as possible on the basis of the global monitoring network and the interpolation by an atmospheric model. The monitoring data is interpolated for the whole globe in the three spatial dimensions and in time, and for all the atmospheric variables needed to initialise the atmospheric model. The atmospheric model is run at a high resolution to provide the deterministic ("best guess") forecast.

In Ensemble Prediction Systems the probability distribution of the future atmospheric state is estimated by running the physically based atmospheric model repeatedly (e.g. 51 times at the ECMWF), each time with a different initial state of the atmosphere. The first run is with the initial conditions as they are assessed for the deterministic run. The spatial resolution of the model used for the ensemble prediction is often lower to reduce computational costs. For the other runs, to make up the probability distribution, the original initial conditions need to be perturbed (changed) in such a way that the resulting forecasted states of the atmosphere by each individual run are equally likely to occur. Together the forecasted states should form a reliable estimate of the probability distribution of what the actual state of the atmosphere will become every time step of the prediction.

The method for preparing the perturbations is different for different institutes. It is often based on statistical knowledge and assumptions of distributions of monitoring and initial state errors, and most importantly, on dynamic estimates of analysis error and error growth. At the ECMWF, for example, samples of possible initial state errors are fed to a simplified, lower resolution, atmospheric model for a limited forecast horizon (36 hours), to find iteratively for which perturbations the development of the atmosphere, in different locations, is changing the most. These paths of rapid changing atmospheric states are called Singular Vectors, and hence, the approach used by ECMWF for its EPS is called the Singular Vector approach (Mureau et al., 1993; Molteni and Buizza, 1996). NCEP uses a method called Breeding. This method is based on consecutive initial state analyses. The biggest deviations between two consecutive initial state analyses indicate areas (locations) of large analysis errors and/or fast growth of analysis errors (Molteni and Buizza, 1996). By choosing the perturbations, such that the potential rapid changes in different locations and directions are covered globally, it is aimed to estimate the (full) probability distribution of future atmospheric states.

The perturbed initial conditions are fed to the global circulation model used to make the final ensemble prediction. Each of the forecasts is called an ensemble member. The result is a number of time series for each of the surface cells of the global model grid, for all atmospheric variables of the model. The medium range EPS of ECMWF and NCEP run from 0 to 15 days ahead. In addition to the perturbations in the initial conditions, also different parameterisations of the atmospheric model are used to account for the model uncertainty. The EPS systems for the medium range from Canada, the USA, and Europe, have been compared in Buizza et al. (2005).

Meteorological institutes like NCEP, ECMWF and the Meteorological Service of Canada (MSC), provide these ensemble forecasts on an operational basis to national and regional meteorological and hydrological organizations (Buizza et al., 2005). These provide Water Boards with access to consistently generated, near real-time, uncertainty information of weather forecasts and the means to generate ensemble based hydrological forecasts. This permits risk based decision making, which has been shown to be more cost effective compared to decisions based only on a deterministic (or single) forecast (Roulin, 2007).

In the Netherlands, the Water Boards are increasingly incorporating the EPS forecasts of the ECMWF into their Decision Support Systems (DSS). In a cooperative project, the Royal Netherlands Meteorological Institute (KNMI) provides warnings for critical events in probabilistic form to several Water Boards (Kok and Vogelezang, 2006, personal communication). The quality of these forecasts and warnings for the Water Boards is still under investigation. On European scale the ECMWF EPS forecasts are used for flood forecasting and early warning for the main rivers in the European Flood Alert System (Werner et al., 2005; Thielen et al., 2009) Long-term verification analysis is needed to develop and test decision rules and control strategies when using the EPS forecasts (Franz et al., 2005).

So far, most studies with ensemble weather forecasts in water management applications have not been concerned with water-system control, but have focused instead on flood forecasting and early warning. Until now, these studies have mostly been performed on single (flood) events (Roo et al., 2003; Bálint et al., 2005; Hlavcova et al., 2005). Only a few studies have carried out a verification analysis for flood forecasting, based on ensemble precipitation forecasts, over a long period of time (e.g. Roulin and Vannitsem, 2005). The present dissertation presents long period verification analyses for water-system control and flood early warning for a water system in the Netherlands and a sub-catchment of the Blue-Nile in Ethiopia.

Next to the ability to "forecast forecast accuracy", a second important benefit of EPS is the increased ability to forecast critical events. Because of the perturbations in the initial conditions, a greater part of the possible spectrum is represented. Therefore, extreme precipitation events will be forecasted more frequent, at least by some of the members.

State of the art research meteorologists aims at verifying the probability of the EPS forecasts of extreme events (Sattler and Feddersen, 2003; Legg and Mylne, 2003; Atger, 2001). These research efforts to provide reliable ensemble meteorological predictions are discussed in combination with related research on hydrological predictions in Section 2.4.3.

2.3.5 Challenges in using weather forecasts for water management

Institutional legacy and confidence
Rayner et al. (2005) have identified several challenges for water managers in using weather forecasts. Institutional limitations consist of a legacy of many years of comparatively successful operational strategies, and of supporting policies and regulations. There is a natural reluctance to change. Even if official policy is changed, the lack of personnel who have been educated in

and are experienced with the new operational strategy limits and delays a shift in practice. These hurdles would however be overcome if (most of) the policy makers and operators were convinced of the effectiveness of weather forecasts and water-system control models. Rayner et al. (2005) have also identified that the policy makers and operators in general remain to be convinced. It does not matter if this is due to unjustified conservatism or justified recognition of the limitations of the modelling systems, because in either case there is still a need for customized analysis methods for long-term verification of forecasting systems and decision rules for anticipatory control actions.

Long-term verification analysis is needed for two reasons. First, verification analysis is required in order to assess the quality of the forecasts for the particular water system at hand. If this is satisfactory, the analysis builds the confidence of the operational water managers to use the forecasts in their decision-making. Second, verification analysis is also needed to develop and test decision rules and control strategies, given the forecasts. In particular, the verification analyses permit decision rules to be simulated so that water managers can see the effect of potential management strategies.

Handling uncertainty: probability forecasts
Another challenge is the analysis, communication and handling of the uncertainty of the weather forecasts. Krzysztofowicz (2001) describes the danger of providing deterministic, single forecasts to decision makers. If such a forecast is considered by the decision maker as representing the "truth", it could lead to disaster. On the other hand, if the decision maker realizes that the forecast is uncertain, but he has no information about the degree of uncertainty, the decision maker may choose to ignore the forecast and delay the decision until measurements come in.

Ensemble prediction systems provide the necessary information on forecast uncertainty. The challenge addressed in the present dissertation is to handle the estimated uncertainty as presented by EPS in decision making for AWM. Simulation models of controlled water systems (Section 2.4) are important decision support tools to transform the ensemble weather forecasts into predictions of water system state, and to analyse the effect of different management strategies.

2.4 Modelling controlled water systems

2.4.1 Definitions
Water systems are continuously becoming more controlled and less natural. Therefore this section addresses the modelling of (partly) controlled water systems and the differences with modelling natural water systems. The focus

in this research is on instantiating a reliable model of a particular water system (Chu and Steinman, 2009), not on hydrodynamic and numerical challenges for computational modelling software (Holly and Merkley, 1993; Clemmens et al., 2005).

Controlled water systems are systems in which the target variables, or the system state, is determined to a large extent by control structures, and not by hydrological processes alone. Only when the hydrological processes go beyond the control capacity of the system, is the system state mainly determined by hydrological processes. An irrigation system is a typical example of a controlled water system. A river is an example of a natural system. A hydropower reservoir is an intermediate example, where the influence of the natural forced river inflow is often large compared to the controllability provided by the discharge structures in the dam. In this research we refer to the first type, the "fully" controlled water systems, like irrigation systems and land-reclamation systems (polders). Many studies focus either on modelling the natural processes, e.g. river flow forecasting problems, or on modelling (optimising) the control strategy, e.g. water allocation problems. The water system and its control strategy need integrated (conjunctive) modelling (Belaineh et al., 1999; Park et al., 2007).

2.4.2 Model components

Models of controlled water systems generaly consist of three main components:
- Rainfall-runoff from contributing catchments;
- Hydrodynamics of flow in conveyance system;
- Control structures and their operation.

The rainfall-runoff part is included, because except for separated (urban) water distribution systems, all controlled water systems still have rainfall as input to their system. The water becomes controllable, only after the rainfall-runoff process has taken place and the hydrological load is being concentrated in the conveyance system, consisting of conduits like sewer pipes and irrigation or drainage canals. The hydrodynamics model the pressures, water levels and discharge in these conduits. In controlled water systems, the up- and downstream boundary conditions are governed by the control structures, as compared to natural boundary conditions like sea-level in natural systems.

The control structures, like pumps, weirs and gates, usually transport water from an up-steam conduit to a downstream conduit or boundary. The discharge through the control structures depends on the discharge capacity of the structure, the operational strategy applied (on-off), the up-stream conduit boundary condition and in some cases the down-stream conduit boundary condition (e.g. head-dependent pumping stations). Next to these three components, increasingly, biological and chemical model components are added and integrated with the hydrodynamics to requirements for (ecological) water quality control (Nestler et al. 2005).

2.4.3 Water system state prediction

The physical processes that govern the state of the atmosphere and water quantity take place at short time and space distances, such that generally Eulerian modelling frameworks (Nestler et al., 2005) are used in this study. In the meteorological forecasting the exception is the application of tracking models for the prediction of path and travel velocity of depressions or storms on the basis of radar or satellite images.

Mostly the target variables are discharge or water level. Discharge is a target variable in an irrigation system, to meet the required volume of irrigation water. Water level is usually the target variable in drainage systems in land-reclamation areas, where water level control is needed to prevent flooding, drought and soil subsidence. Water quality has long been a secondary target variable in irrigations systems, where flushing is needed to prevent salinisation of the soil. Water quality control is becoming more important, because of ongoing urbanisation and subsequent increasing pressure on waste water treatment plants and receiving waters.

The water system state predictions in this research concern the hydro-meteorological ensemble forecasts. Next to the uncertainty in the precipitation forecast as expressed by the meteorological EPS, other sources of uncertainty in hydro-meteorological forecasts are: parameter uncertainty, model structure uncertainty, initial state (measured, data assimilation) uncertainty, and statistical uncertainty (Maskey, 2004; De Vriend, 2002). The aim of ongoing research in hydro-meteorological ensemble forecasting, is to develop methods such that the forecast uncertainty is well represented (e.g. GLUE method: Beven and Freer, 2001), while at the same time uncertainty is minimised as much as possible in a Bayesian approach (Krzysztofowicz, 2002; Todini, 1999). A scheme, generally adopted in the international Hydrological Ensemble Prediction Experiment (HEPEX), of a general hydro-meteorological EPS (Schaake et al., 2007) that encompasses these two characteristics is presented in a simplified manner in Figure 2.1.

The meteorological pre-processor stands for all analyses that make the ensemble weather forecasts more suitable for use as input to a hydrological model. More suitable means, for example, reliable probability forecasts and sufficient spatial resolution. Despite the efforts of meteorological institutes in preparing the ensemble forecasts, for particular variables at particular locations systematic errors often occur. On the basis of archived local monitoring data and archived forecasts or hindcasts, bias correction can be performed (e.g. Clark et al., 2004). For processing the weather forecast output to the required resolution, downscaling techniques need to be applied. Downscaling techniques can be divided in statistical downscaling, dynamic downscaling, and analogue methods (Hamill, 2009). Statistical downscaling uses similar bias correction techniques relating high resolution monitored data of the target variable with the forecasted data. Dynamic downscaling is done by nesting finer resolution atmospheric models in the atmospheric model used for the ensemble prediction system. Analogues refers to pre-processing weather forecasts on the basis of comparison with historically similar forecast and actual state characteristics. The hydrologic post-processor performs tasks similar to the meteorological pre-processor, but now on the basis of historic and real-time data of target water system variables (e.g. Olsson and Lindstrom, 2008).

In this dissertation focus is on the end-use of ensemble prediction. This step requires the interpretation of the ensemble prediction for decision-making. In Section 2.5 it is suggested to expand the hydro-meteorological EPS scheme of Figure 2.1 with, what could be defined as a "Decision support pre-processor".

Next to the sources of meteorological and hydrological uncertainty, in applications of Anticipatory control also the uncertainty in control actions and system response needs to be taken into account. The control uncertainty is discussed in Sections 2.4.4.

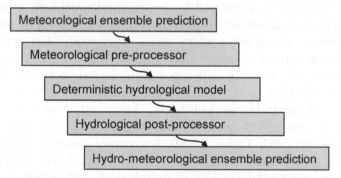

Figure 2.1 Elements of a hydro-meteorological ensemble prediction system

2.4.4 Challenges in modelling controlled water systems

A challenge of modelling controlled water systems is that the modelling problem often consists of a high degree of freedom. For example, a model of a controlled water system with water level as the target variable can be considered, e.g. a drainage canal. The water level in the water conduits is to be controlled within a pre-defined range by a down-stream pumping station. Then, the water level can easily be modelled by including a pumping station with switch on and -off levels according to the target control range. Even without the correct rainfall input and discharge capacity of the pumping station, the model would produce a fair reproduction of the actual water level, because the modelled pump would simply adjust its pumping frequency and duration to keep the water level within the control range. The target variable, water level, is then modelled accurately, but the operation of the pumping station, and the volume that goes through the system might be totally different from reality. This illustrates the extra degree of freedom that is created by the inclusion of control structures in models. The risk of producing the right output for the wrong reasons when modelling controlled water systems, therefore, does not only refer to the risk of over-calibration but also to the risk of providing nonsense system characteristics and input to the model without noticing directly in the model results. Likewise, the actual control structures in the actual water system also give a high degree of freedom to the operators on how to manage the water levels. This can make the water levels highly unpredictable, because operators may depart from operational routine at any moment. This makes it more difficult to model controlled systems accurately for long time series, compared to natural systems.

An advantage of modelling controlled water systems is that in many cases the data availability is better than in natural water systems. Regulating structures usually perform discharge measurement at the same time, and in addition up- and downstream measurement stations are often installed. The cross-section is usually known for pipe-networks, canals, and well-maintained ditches, with higher accuracy than the cross-sections for natural streams. Although for the underground pipe-networks this is only true when proper reporting during construction has been performed.

These particular issues in the modelling of controlled water systems are used in the formulation of a modelling framework in Section 3.4.2.

Another challenge, particular for ensemble hydro-meteorological forecasting is the time constraints involved in applying all ensemble weather forecast input to the hydrological model. This is sometimes solved through aggregating the ensembles by taking percentiles. However, this leads to time series that have never been predicted, and as such may not make any sense deterministically. Selection of a number of ensemble members always runs the risk of missing extremes. Application of fast models or high performance computing would allow inclusion of all the ensemble information.

2.5 Decision making with uncertainty

2.5.1 Uncertainty

The different sources of uncertainty in weather forecasts and water system state prediction have been mentioned in Sections 2.3 and 2.4. The problem for decision making of the uncertainty of forecasts is two-fold. First in an absolute sense, both the weather forecast and the system response modelling cannot be fully accurate, especially when the uncertainty in the weather forecasts is high. Secondly, the uncertainty is difficult to assess beforehand. A decision maker would benefit from knowing the probability that the forecasted event will actually occur. Only then can he make a cost-benefit based risk analysis of taking or not taking anticipatory measures.

In water management decisions, three forms of inaccuracy are important. If we want to optimise water management with the advent of an extreme event, the event has to be forecasted correctly in terms of:
-    location
-    timing
-    magnitude

If the event occurs in a different location, outside the water management system of our interest, anticipatory measures are in vain. If the event arrives sooner than expected, anticipatory measures will not have enough time to become effective and safety is at stake. If the event arrives later the negative consequences depend on the degree of optimisation of the system. It could be argued that the more a system is optimised, the greater the damage when the system fails. An emergency retention basin, for instance, needs a perfect forecast of the timing of the flood wave. If the flood wave arrives later than expected, the filling of the retention basin starts too early and by the time the peak of the wave arrives, the basin is already full and flooding will occur. Examples of forecast magnitude are the precipitation depth, hydrological load or water levels. Thus, the sensitivity to particular forms of inaccuracies depends on the anticipatory management action.

Unfortunately the uncertainty of forecasts is neither constant in time nor uniform in space. Weather above large flat land areas can be forecasted better than weather in mountainous areas. The trajectory of large frontal systems is sometimes easier to predict than the inception of convective storms. Wet season forecasts might generally be more accurate than dry season forecasts. The same applies to water system models. Some models may be tuned to perform well during average hydrological conditions, others on critical events. A model of a system with one straight concrete lined canal and a weir maybe more accurate than a model of a complex network of canals with peat embankments. Therefore, generalised performance indices on a regional or continental scale, as provided by meteorological institutes, give the water authorities only little information. Water authorities should perform verification analysis for their own water system, and customised for the type of Anticipatory Water Management to be applied. In this way the water authority can train (calibrate) its interpretation of the ensemble hydro-meteorological predictions for decision support in AWM (Decision support pre-processor, Figure 2.2).

For assessing the behaviour of the Ensemble Prediction Systems to improve decision-making, verification data is needed. In many cases verification data will be limited, since the safety level is already high and thus extreme events of interest seldom occur. On top of this the meteorological forecasting models are being continuously developed, which reduces the number of extreme events to verify the forecasts.

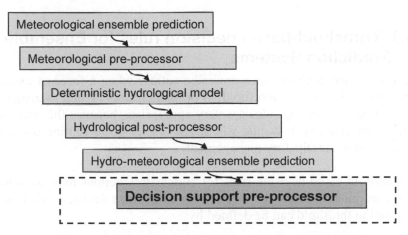

Figure 2.2 Elements of a hydro-meteorological ensemble prediction system expanded with a Decision support pre-processor for end-use of the predictions in Anticipatory Water Management

2.5.2 Risk

It has been discussed that hydro-meteorological forecasts are uncertain. This would not be a problem if there would not be any risk of taking anticipatory actions based on these forecasts. In water management problems, however, taking unnecessary actions or not taking action when it is necessary usually has negative implications. Meteorologists speak of false alarms and misses.

In the case of a false alarm economic damage will often occur. Evacuations, releasing water from hydropower reservoirs and operating pumping stations cost money. But also safety could be endangered. Imagine that a false alarm of an intensive precipitation event is followed by anticipatory lowering of reservoir levels below threshold values safeguarding the stability of embankments. The forecasted precipitation is expected to set up the water levels soon after. If the precipitation event does not come, the water levels will stay at this dangerous low level.

Misses of extreme events could be even more dangerous. If the flood defence strategy of a water system relies on the forecast of intensive precipitation events, missing such an event poses serious threats to the community. If policy guidelines on safety have to be met by Anticipatory Water Management, the frequency of forecasts missing extreme events should be very low.

These risks in terms of safety and damage (economic, social, nature, etc.) have to be assessed and taken into account when deciding on which levels of uncertainty are acceptable.

2.5.3 Threshold-based decision rules for Ensemble Prediction Systems

Threshold-based decision rules prescribe actions when a forecast exceeds a predefined value (event threshold). For Anticipatory Water Management, threshold based decision rules are very appropriate, because the first step in the decision chain is to decide whether to switch from routine operational management to anticipatory management.

The threshold-based decision rules for anticipatory actions may consider the use of precipitation forecasts directly. For example, a decision rule based on precipitation threshold can be defined by:

If forecasted precipitation > X mm day-1, then start control action A.

This decision rule is suited for a deterministic, single precipitation forecast for a fixed 1-day forecast horizon. In the case of ensemble prediction systems, there are a number of possible forecasts to consider for a range of forecast horizons. The more of the ensemble members that exceed the precipitation threshold, the higher the forecasted probability that the precipitation threshold will be exceeded. A decision rule based on EPS has to define how high this forecasted probability (P) should be, before an anticipatory action is taken (see the probability threshold in Figure 2.3):

If forecasted probability P(precipitation Y days from now > X mm d-1) > N, then start anticipatory action A.

Water Boards often have no information on how to set these decision rules. The effects of the choice of event threshold, forecast horizon and probability threshold on the performance of the management strategies are not known.

There is no knowledge about the performance of the particular weather forecasting system (e.g. ECMWF EPS), for the given water system, let alone information about the performance of the decision rules that depend on these forecasts.

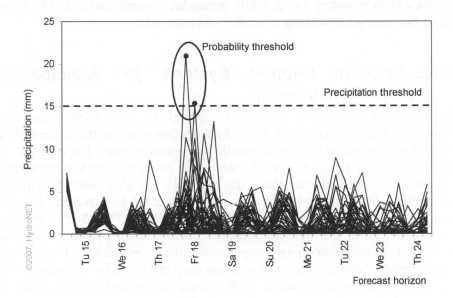

Figure 2.3 ECMWF EPS precipitation time series for location De Bilt (NL) (data source: KNMI). When applying threshold-based decision rules for EPS, the event threshold (Precipitation threshold), the forecast horizon and the probability threshold have to be set. The probability threshold is the required forecasted probability that the precipitation threshold will be exceeded. This is determined by the ensemble members exceeding the precipitation threshold.

2.5.4 Cost-benefit analysis

The choice of a particular decision rule will not only depend on the accuracy of the decisions. A cost-loss analysis is needed to optimize further the decision rules and control strategies. The cost of an anticipatory control-action and the loss if no action is taken when a critical event occurs are different for every application, every particular water system, and for every event. Furthermore, the anticipatory action does not have to be fixed, but may be optimized for every particular forecast.

The cost-benefit analysis needs to take into account different types of damage. The direct damage is the damage that occurs immediately after a flood event due to the physical contact of the water with humans and damageable properties (Smith and Ward, 1998). Indirect costs accumulate over time after the event, for example because the damaged infrastructure is reducing economic output in the affected area. Next to the distinction between direct an indirect costs, the differences between tangible and intangible damage have to be taken into account. Tangible damage is referred to in this dissertation as damage that can be expressed (and estimated) in monetary units, while intangible damage can not. The most important intangible damage is the loss of human life.

2.5.5 Decision Support Systems for Anticipatory Water Management

Incorporating probabilistic forecasts of several days ahead into a decision support system (DSS) brings along high requirements of the system. The DSS-system needs to be able to handle many data streams. For example, if we get a 10-day forecast every 12 hours, we could save each forecast and compare 14 forecasts for three days ahead. We also want to have a probability distribution of each forecast, which could be represented by 50 possible forecasts of the same time frame (ECMWF EPS). Add to this the measured data in the field for update or assimilation purposes with the system response modelling output, and the complexity of the data management becomes clear. The decision support system has to channel this data through all its components, such as database, system response models, optimisation models and presentation modules.

The optimisation problem also becomes very complex. Because anticipatory management involves risks, it is preferred to stick to normal management. Therefore, first the system has to be optimised using the constraints of normal management. If normal management is expected to fail after, for instance, four days (with enough probability), anticipatory actions will be applied and then has to be optimised as well. This again is constrained optimisation, e.g. to the system safety limits and capacity. At the next decision moment both optimisations have to be applied again. Finally, uncertainty must be handled as well, either in the objective functions or in the presented results. The water manager can add to these results his expert knowledge to complete the risk analyses and make the decision.

Complexity does not have to be a problem if there is ample time to solve the optimisation problem. But water systems are dynamic and optimisation of operational management therefore requires frequent decision moments. Furthermore, anticipatory actions take time before becoming effective and the lead-time of forecasts is limited. Thus the computational speed of decision support systems could become a limiting factor.

The last issue to be raised on decision support systems is their reliability. This research seeks to apply anticipatory management to reduce flooding problems. Thus the reliability of the entire system has to be high. Anticipatory management will depend strongly on communication lines for collecting the data and forwarding control decisions to operators. These communication lines have to be secure. Another well-known problem is the reliability of the computational models. Run-time errors are annoying for analysis purposes, but very harmful for real-time control applications.

Note that in this section on DSS it is implicitly assumed that the decision problem is mostly too complex to be solved by managers without the aid of state-of-the-art support techniques. This assumption has to be treated carefully. Many water managers have years of experience in operating the control structure(s) of their water system. They also have considerable knowledge on the system response to weather events. Thus anticipatory management could be done by, for example, just providing weather forecasts. The system response model will then be the experience based models of the water managers and decisions will be based on their risk analysis and priorities.

In practise many of the developed and operational decision support systems are only partially used in the actual management of water systems. The role of decision support systems is not formalised in management regulations. Sometimes systems are not reliable enough or not practical to use (too slow or too complicated). Most importantly they are not (and cannot be) 100% accurate, while their deterministic outcomes suggest the opposite. Many water managers, therefore, put little trust in weather forecasts and decision support systems.

2.6 Knowledge gaps and hypotheses

To summarise, the main challenges of expanding the decision horizon by the use of weather forecasts in Anticipatory Water Management are the high uncertainty of the future system state and the risks associated with in-appropriate anticipatory management actions.

From the literature review it appears that there are opportunities to enhance AWM through the availability of ensemble weather prediction systems. However there is a first knowledge gap on how the ensemble weather forecasts perform over a long period of time for a particular catchment. There are no universal guidelines on how to assess this. The hypotheses are that:

> The comparison of measured precipitation local to a given water system, with ensemble precipitation forecasts leads to an improvement in the use of those forecasts. (Hypothesis 1)

and that:

> The comparison of measured water levels in a system, with those predicted in response to ensemble precipitation forecasts under the current management strategy, leads to an improvement in the use of that system. (Hypothesis 2)

The second knowledge gap is on how to define the best decision rules with these probabilistic forecasts for whether to anticipate or not. It is hypothesised that:

> Effective decision rules can be found by hindcast analysis. (Hypothesis 3)

Once the warning that anticipation is necessary has been given, the third major knowledge gap is on what exactly the Anticipatory Water Management action should be. It is hypothesised that:

> Long-term simulation of the complete Antcipatory Water Management strategy for historic time series enables an optimisation of AWM. (Hypothesis 4)

Then, with the AWM strategy developed it should be decided by the water authorities whether the strategy should be applied. It is hypothesised that:

> A cost-benefit analysis, based on the continuous simulation of water levels, generated in a water system, with a prescribed management strategy, by forecasted precipitation from a specific product for a historic time series with critical events, is needed to assess whether the forecasting product should be applied to the particular system. (Hypothesis 5)

This implies that the costs and benefits of inappropriate and appropriate anticipatory management actions have to be assessed dynamically, and an adequate simulation model of the controlled water system has to be available. These two requirements are often not readily available for a particular water system, or with a particular water management body, e.g. a Water Board.

A framework for developing Anticipatory Water Management for a particular water system is needed to fill these interrelated knowledge gaps and requirements. This framework is prepared in the next chapter. It is hypothesised that when implementing this framework:

> The use of ensemble precipitation forecasts to decide on anticipatory control actions, in preference to re-active control, can reduce the damage over a long period of time. (Hypothesis 6)

and that:

> The expected benefits when applying AWM, despite their uncertainty (due to limited availability of data, changes in the cost-benefit relationships, model uncertainty, etc.), more than compensate for the losses that occur when AWM is not applied. (Hypothesis 7)

3 Framework for developing Anticipatory Water Management (AWM)

3.1 Introduction

This chapter discusses the different steps, and the associated challenges, for developing and evaluating an Anticipatory Water Management strategy for a particular water system. The resulting theoretical framework is put into practice in two cases studies described in Chapters 4 and 5.

3.2 Establishing the need and potential for AWM

3.2.1 For which events is AWM needed

Before exploring the possibilities of AWM it should be clear why and for which events AWM is needed. Although this seems an obvious first step, there are a number of reasons why this is often not taken and why it is not straightforward to complete this task. Let us first consider why event selection needs to be done and which criteria the event selection should adopt. As in any scientific study, engineering project and management review, there should be an incentive to change the present situation. For all AWM applications the incentive to change is universal in the sense that there is a desire to act on upcoming events, before they actually happen. This means that from the past history or future scenario outlooks, it should be known which events led (or will lead) to undesirable situations (damage). For flood control this refers to past flood events. While recognising that not all extreme events can be handled by improved control, AWM can potentially reduce the frequency and magnitude of the damage due to these events. In order to develop decision rules and to evaluate the effectiveness of AWM as many events as possible should be identified. The higher the number of selected events the better the confidence in the analysis results. For added confidence in the findings, it is therefore also preferable to have historic events, instead of hypothetical events.

In scientific studies, such as hydrological modelling, the event selection is often not done. The focus is on reproducing (characteristics of) hydrographs. Often continuous simulation is applied and the performance is measured in terms of standard values, such as the squared error, in order to be able to compare with other models and earlier publications. In the case that the intended end-application is flood forecasting, emphasis is placed on the peaks of these hydrographs; but still it is often not discussed exactly which peaks at what moment led to flooding historically. Then in the modelling stage the model is calibrated and validated, and subsequently presented to the end-user. The end-user usually does have some idea of which events are critical for his water system, and based on forecasts of these he could make warning decisions, but these end-user decision rules are rarely evaluated for a series of past events.

Methods for event selection
The interesting thing is that if past events are selected, this appears not to be as easy as expected. In many countries secondary data on flood events is limited, or there is limited access. The latter arises either because of political sensitivity or because of lack of (digital) infrastructure to search for this data. The data can be limited in the sense that date of flooding may be known from old newspapers, but the exact location of affected areas, the cause and the duration of the flooding are not. This is particular truth for poor countries, as is discussed in the Ethiopia case study (Chapter 5).

Secondly, in controlled water systems it is often difficult to determine critical events, because of the human based regulation of the structure. For instance when looking at water level records of a small reservoir, high peaks do not necessarily mean that an extreme natural event occurred. The high water levels may also have been caused by "wrong" operation of a discharge structure (Figure 3.1a). Vice versa, an extreme load to the reservoir may not be observed by looking for peaks in the water level data, because the reservoir level was unusually low at the beginning of the event (Figure 3.1b). This can be because release was maintained too long after a previous event, or because a cautious operator decided to apply an anticipatory release himself. In the following, solutions for both lack of data and deceptive data are given.

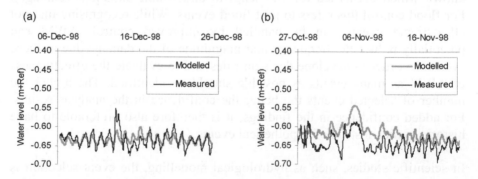

Figure 3.1 Unnecessary high measured water levels (a) and high measured water levels prevented by early lowering of storage level (b)

Finding system thresholds in data scarce catchments
Anticipatory Water Management is necessary when the capacity of the water system is not sufficient to cope with the hydrological load. Every water system has limits in the amount of water that it can discharge and store, meaning that up to the point where this limit is reached the system functions well, and when the limits are exceeded and too much water is entering the system, flooding will occur. On the other hand, every water system has limits to the amount of water it can let in, recharge or store, meaning that if for too long a period the net inflow is negative, water levels will drop and drought may occur. To know whether Anticipatory Water Management should be applied, therefore, involves finding these system thresholds. Interviews of inhabitants of flood prone areas, government and NGO staff members, and water professionals help to determine the timing and affected area of the events.

In data scarce countries and catchments the first step is to look into secondary data like (electronic) newspapers and humanitarian organisation reports. This results in at least the years when floods and droughts occurred. Most of the time, it is also possible to find from these sources the start of the event with several days accuracy. Some idea of the severity of the event and the cause may also be given. In these information sources, affected areas are usually only roughly mentioned, like a regional district level, or affected cities and villages.

A better source of information on the spatial impact of an event can be found from satellite images, like radar-sat, which are increasingly available. Next to a precise delineation of the affected areas, these images also provide a confirmation of the timing of the event, because the exact data and time the image was taken is always given. Note that the timing cannot be taken from satellite imaginary alone, because in the case of flooding the affected area may be inundated for several days up to a month. It cannot be seen from the image whether it was taken at the beginning or at the end of the event.

All these secondary data can be compared with the primary data, the time series data. If start dates are known, system thresholds may be found directly if the dates are consistent with the highest (or lowest) recorded peaks. Errors in the data can also be an indication of extreme events. For example flattened peak discharges may result from water levels rising above the measurement scale. Missing data can indicate the moment that a measuring device was flushed away. Note that all this information has to be used with care, because its interpretation can vary; for instance, the stopping of a recording can also just mean that it was damaged by a floating log, or by vandalism. Also, from the time-series data it cannot always be inferred what was the cause of the event. It may be that the capacity of the system was still sufficient, but that system failure led to the damage. In the case of flooding a (intentional) breach of the embankments may be the cause; in the case of drought, illegal extractions may be the cause.

Also, the variable in which the threshold is expressed can not often be decided upon beforehand. While in reservoir systems water level is often chosen, precipitation depth is some cases may be used as well (avoiding the dependency of rainfall-runoff, hydrodynamic and reservoir modelling), or inflow volumes etc. A joint use of different variables can be used to cover a wider range of forecast horizons and to cross-verify forecasts.

It can be that the available information is not sufficient to determine the system thresholds. In that case it must be asserted what information is trusted and to what detail the occurrence of extreme events can be determined. For instance, it may be possible to assess only the years in which flooding results. Then a range of thresholds can be estimated, for instance by comparing the hydrographs of the years with flooding to the years without flooding. An example three-level threshold system consists of an alert threshold (AWM may be necessary), alarm level (AWM must be applied), and disaster level (AWM will not help, calamity plans must be executed).

Finding system thresholds in controlled water systems
In controlled water systems more data are often available. In the channelled reservoir systems used in the Netherlands for draining the low-lying reclaimed lands called "polders", precipitation data, water level recordings, and regulating structures operation and discharges are usually available for several years. The problem comes from the not fully consistent, human-based, operational management of the storage basin water level. Here a thorough system analysis, taking into account all relevant variables must be applied.

The primary variable from the systems approach for expressing thresholds is Volume. This volume, however, is usually not directly measured or measurable. In the Netherlands for instance the discharge of the smaller pumping stations from the low-lying areas to the higher channelled storage basin is not always available. Also, especially for the longer time scales as with drought, seepage, percolation and evaporation become relevant, and yet are difficult to measure. Therefore, it is often preferred to use the resultant variable, water level, which is easy to measure. Water level is often the primary variable from the point of view of impact on society, because it is the high or low (ground)water levels that cause flooding and drought related damage. From the system design it is already known what the water level thresholds are. The historic water level recordings that exceed these thresholds should be identified as a first step to identify water system thresholds. In step two the selected events should be filtered by looking at the precipitation data, the regulating discharge data and other available volume data to exclude the extreme water level events that were caused by the control strategy or regulating structure failure, instead of exceedance of the systems capacity. In the third and final stage, events should be added that did not exceed the water level thresholds because a temporarily change of control strategy was applied. For instance, lowering reservoir levels before an extreme event occurs may prevent the water levels from exceeding the threshold, although the inflow volume was more than the system's capacity. Note that in this case, Anticipatory Water Management is already successfully applied, even before it is formalised in the operational policy and control strategies. These cases are not uncommon, because operators have many years of experience and have a mental model that includes system thresholds and they are used to dealing with additional information like weather forecasts from the television. These events can be identified by checking water level records for unusually low water levels just before an unusually long rise of the water level.

After the critical events have been selected, based mainly on the resultant variable, water level, the analysis can be expanded to find the volume based system capacity as well. If, as was described in the beginning of this section, discharge volume data is not available (small pumping stations in polders), then the volume analysis can be taken further upstream in the hydrological cycle, e.g. up to the precipitation input volume. Analysis of the relationship between measured precipitation and extreme events can very well serve as a cross-validation of event selection in controlled water systems or as additional guidance for anticipation (Figure 3.2).

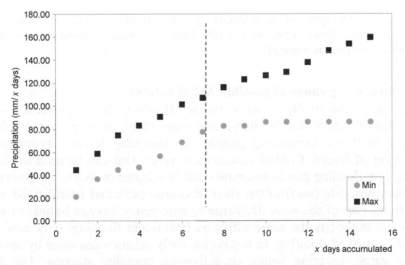

Figure 3.2 Upper and lower precipitation thresholds for accumulated precipitation in Rijnland. After seven days (veritical line) the minimum threshold does not increase anymore.

3.2.2 Potential for anticipatory management action

After the critical events have been clearly defined in terms of thresholds, the Anticipatory Water Management actions that may reduce the (frequency of) exceedance of these thresholds have to be defined. Examples of such actions are maintaining reservoir levels in anticipation of a dry-spell, and lowering reservoir levels in anticipation of a peak discharge (Figure 3.3). Conceptualising the pro-active actions can best be done by or together with the operational managers and policy developers of the water authority responsible for the particular water system.

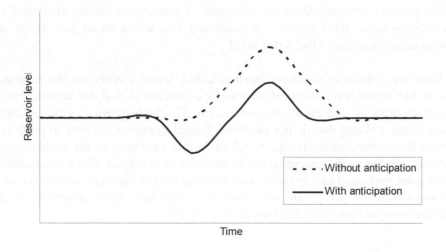

Figure 3.3. Example of anticipatory action. Reservoir level is lowered in anticipation of a flood event. As a result of the anticipatory lowering, the resulting peak reservoir level is reduced.

Conceptualising range of possible AWM actions
In natural uncontrolled rivers where flooding is a problem, water management actions can be; harvesting crops early, moving assets to higher ground or floors, evacuating people, strengthening levees, or intentional breaching of levees. Control structures in rivers that can be used to reduce damage by flooding can be reservoirs and floodgates or weirs. The reservoirs can accommodate (part) of the river discharge peak and lateral flood weirs can divert part of the peak discharge to emergency storage basins. In some cases the latter has the same effect as intentional flooding, only now it is called controlled flooding. In reservoirs, early releases can done by opening sluice gates, lowering weirs, or activating pumping stations. The early releases lower the water level in the storage basin, thus increasing the storage capacity for the upcoming peak inflow, or hydrological load. Anticipatory actions for events invoking a shortage of water, range from crop cultivation strategies, via water supply rationing, to keeping reservoirs

full at the maximum level. From these examples it can be seen directly that there is a wide variation in the time these actions take to become effective. Therefore, after having an anticipatory control action in mind, the entire process to apply this action has to be conceptualised in order to determine the required forecast horizon.

Conceptualising the AWM process and estimating required forecast horizon
Whereas there is a great variety of AWM actions and applications, the general process is always the same (Figure 3.4): forecast, communicate the forecast to a decision maker, choose anticipatory action, communicate the decision to operators, implement anticipatory action, allow time for the action to become effective (action response time).

Every management action takes time to become effective. Evacuation takes hours to days or maybe even weeks to be completed. Sluices have to be opened before water can be released from a reservoir. The response time of the water levels in the reservoir and the upstream water system depends on the discharge capacity.

Next to this action response time, the decision-making process also takes time. The information that is used by the decision makers and decision support systems is not in real time. Measurements have to be communicated from the measurement station to the central control unit (often to a database), where it might have to be processed before it can be used. Decision makers have to interpret the information that is given to them. Hydrological models, control models and optimisation models need time to run. In many cases decisions have to be deliberated amongst several actors. Also the communication of the decision to operators takes time.

Weather forecasts and analyses are made at the meteorological institutes and it takes a considerable time before the forecasts are available for water managers. In the Netherlands the delivery delay of the 10-day ensemble forecasts of the ECMWF is about 15 hours.

The management action response time, decision time and data delivery delays need to be taken into account when determining the required forecast lead-time. Note that although the timing of these processes can be estimated from a knowledge of the water system and experience with the operation of the regulating structures, in many cases at this stage a computational model of the water system is needed, because control actions are being considered that have not been (regularly) applied in the past. We will return to water system control modelling in Section 3.4. and Chapter 4.

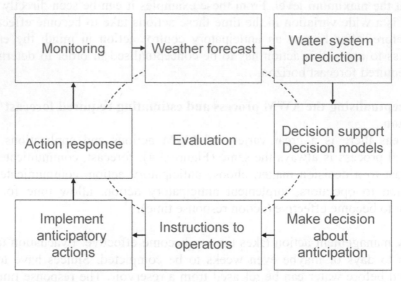

Figure 3.4 General process of anticipatory water management

Setting boundary conditions
Boundary conditions are determined by the system design and its user requirements. Reservoirs levels cannot drop below the dead storage level and because of user requirements, such as power generation, the water level is not allowed to drop below the upper active storage level. In channelled storage basins, upper lower limits are governed by water depth requirements for navigation, bank stability problems and groundwater level requirements for crops.

Determining potential effectiveness of AWM actions
With the boundary conditions for the anticipatory management actions known, the potential effectiveness of the actions can be determined. This can be simple volume calculations with the storage basin level-area function and the planned lowering or rising of the level before a flood (or drought event), or model runs where historic extreme events are simulated with the lower (or higher) antecedent storage levels. Because this is to calculate the potential effect of the AWM, these model runs assume perfect forecasts, e.g. only measured inputs are used. In the case of flood and drought forecasting and early warning the effectiveness is often more difficult to quantify. Here the effectiveness depends on agricultural processes, evacuation processes and emergency mitigation measures like putting up temporary levees.

Identifying risks of adverse effects of AWM actions
It is inherent to AWM that proposed control actions deviate from the normal operational guidelines. AWM deals with temporally expanding the normal control range, which increases the risk of adverse effects. These risks have already been taken into account in conceptualising the anticipatory control

action and setting its boundary conditions, however, this is done for the case when the AWM action is applied correctly. Additional risk analysis should be performed for the cases that the AWM is not performed correctly. If AWM is considered as a measure to replace structural measures such as an increase in storage capacity, than the first risk is that an extreme event is not forecasted and the AWM actions are not implemented or are started too late. This leads to increased risk of flooding (or drought). The second risk is that of "false alarms" or in a continuous sense over-predicting the upcoming event; this always needs to be considered. Forecasts are never 100% certain and because of the increased horizons applied for AWM, higher levels of uncertainty are to be expected. These result in unnecessary (or too aggressive) AWM actions. In the case of flood control in a reservoir this means that the reservoir level is lowered because high inflows have been forecasted, while subsequently the expected inflow does not arrive. This could lead to prolonged low storage levels, with risks of water shortage, falling groundwater levels, failure to meet power generation requirements etc. The risk of false alarms is part of the "potential" of AWM and needs to be taken into account in the development of AWM.

3.3 Verification analysis

3.3.1 Product selection: time scales, spatial scales

When the aim of AWM has been identified, including the range of events for which it should apply and the anticipatory actions that are to be taken, a forecasting system can be selected. The forecasting horizon should be enough for the AWM to be effective. The time step and spatial resolution of the meteorological forecasts should match the water system characteristics.

After the forecasting system has been selected, the uncertainty and associated analyses will have to assess up to which lead-time useful information for management can still be deduced from weather forecasts (Figure 3.5).

3.3.2 Continuous simulation of the real-time AWM forecasting system

The uncertainty of the forecasts is of such importance for the effectiveness of AWM and for the development of decision rules that its assessment must be part of a fast screening analysis. It can be fast because the AWM strategy does not yet have to be worked out. The only focus now is to make forecasts with the current, normal, control strategy and to verify these forecasts with historic data. The verification analysis is a screening of the forecasting product that is being considered for use in AWM.

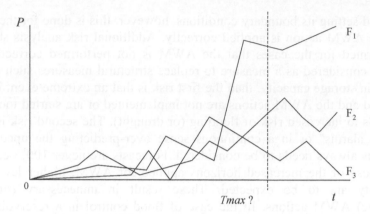

Figure 3.5 Fictitious example of a three-member ensemble precipitation forecast. At a certain lead-time the uncertainty might be considered too high for decision making (Tmax).

This can be a meteorological forecasting product, hydrological, hydrodynamic, or any considered change or addition to the tools used for operation management decisions.

For this screening to be effective the real-time forecasting process that will be used in the AWM needs to be emulated. This approach is called hindcasting. A modelling system is prepared that allows continuous hindcast simulation of long periods (multiple years). The layout of such a modelling system is given in Figure 3.6. The available length of the time series depends on the water system data in the water authority's archive and the forecast or re-forecast archive of the new meteorological forecast product, provided by the meteorological organisation.

First, the water system model used, must be validated for the continuous run using measured input data, instead of archived forecasts. For controlled water systems this is to test whether the normal control strategy can be modelled well. This is a pre-requisite for the hindcasting to make sense. The hindcasts are to predict which event could not be handled with the normal control strategy, and can be handled better with AWM. Once the water system control model performs satisfactorally it can serve to generate the reference time series, instead of the measured time series. This is because the model shows what would have happened if the control strategy had been executed consistently, while the measured data has all the unpredictable, human control, decisions incorporated in it. This water system control model also helps to identify the critical events (Section 3.2.1) that were not visible in the recorded data because accidentally AWM was applied instead of normal control.

Finally, a good water system control model is important to build confidence with the water authorities that the AWM strategies that will be developed and evaluated using this modelling system are realistic. Water system control modelling is discussed separately in Section 3.4.

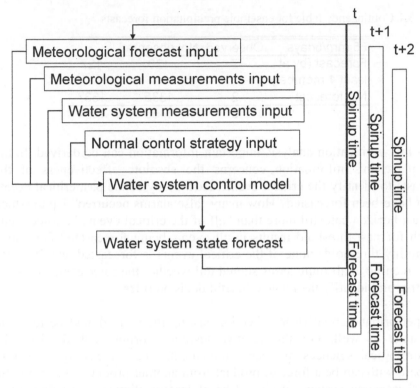

Figure 3.6 Creating hindcasts. The forecasting process is repeated for every time step *t* in the past.

3.3.3 Event based verification of a range of decision rules for AWM

The next step of the screening is to verify the produced hindcasts with the selected critical events. If the forecasts identify the critical events, it means that the forecasting system can be used to decide when anticipatory actions are needed. Note that this is only the first step in an actual AWM strategy, because after the decision is made to undertake pro-active action, it has to be decided what exactly the action will be. Since this is a dichotomous (anticipate 'yes' or 'no') decision problem the verification results can be classified according to a contingency table, where every forecast is either a hit (event occurs and is forecasted), a false alarm (event is forecasted but does not occur) or a correct rejection (event is not forecasted and does not occur) or a miss (event occurs but was not forecasted) (Table 3.1). Note that this is different from the usual verification of hydrological and hydrodynamic models. The verification has to be done for a range for decision rules that are considered for the AWM strategy.

Table 3.1 Contingency table for ensemble precipitation forecasts

65 mm/5days	Observed	Not observed	
Forecast (by at least 1 member)	7	120	127
Not forecast	2	1335	1337
	9	1455	1464

From this verification analysis important information can be derived. In the first place this information concerns the absolute effectiveness of the forecasts to identify the critical events. How many of the past critical events could have been forecasted? How many false alarms occurred? Up to which forecast horizon can still more than half of the critical events be forecasted? Which forecast threshold results in all events being forecasted? For which probability thresholds none of the critical events is forecasted, etc. In short, this screening in absolute terms should tell whether the forecasting system is effective and identify the range of useful decision rules.

The performance relative to other forecasting methods should be tested at this stage as well. For this relative testing, scoring methods from the meteorological sciences are very convenient. The forecasting system to compare with can be a forecast product from another producer, the presently used forecasting system, or should be at least a climate based forecast, a forecast as previous, or a random forecast. The climate forecast is mostly used as a baseline forecasting system. Here the forecasted probability of an event is the climate frequency of the event. This frequency is often assumed to be equal to the sample frequency from the analysis record. The new forecasting system should of course be at least better than the climate forecasts. Relative characteristics of a forecasting system are expressed in "skill scores". For this screening the relative operating curve is often used (Kok, 2000, p. 59) , because it visualises skills and decision rules of different forecast systems in one graph.

These meteorological skill scores are calculated by evaluating each forecast. In Anticipatory Water Management it is better to apply an event based verification approach. The comparison between the measured and forecasted events should be done for the time at which a critical event begins. The forecast of the beginning of an event is important to allow for effective anticipatory control actions. The forecasted time at which the event begins should be within a predefined range (e.g. one day) of the actual beginning of the event.

Note that this is different from a "forecast by forecast" analysis. Events that last more than one day are considered as only one event, so that correctly forecasting one long event does not count as multiple hits. This avoids masking missed events of short duration by correctly forecasting one of the long events. In the same way the missing of one long event, is considered as one missed event. This avoids the disbenefit of correctly identifying several short events whereas only one long event has been missed. Such considerations are particularly important in the analyses for anticipatory water-system control because they deal with infrequent critical events. Also, false alarms are analysed as separate events. The drawback of the event based approach is that forecasted duration of events is not scored, which in some cases may result in less stringent verification than the "forecast by forecast" verification.

3.4 Modelling controlled water systems

It has been made clear that for a successful verification analysis a reliable water system control model is needed. Therefore, in this section, modelling of controlled water systems is discussed.

In Section 2.4 the main challenges in modelling controlled water systems have been identified as the high degree of freedom and the unpredictable human based control strategies. In many cases the choice is made to model only the rainfall-runoff part (Roulin, 2007) or to take a set of control rules and consider the water system control model as providing a potential / or perfect result. The main reason for this is that human behaviour is difficult to predict and hence difficult to model. Most controlled surface water systems have considerable or exclusive human supervised control. Therefore the control is often not fully consistent over time.

Here it is argued that still it should always be attempted to model the current (business as usual) control strategies and to show the model results together with the measured results. The need for modelling control strategies, as compared to modelling only the rainfall runoff process, is clear when the aim is to evaluate new control strategies. The reasons for presenting the comparison with measured data are three-fold:

1. For the scientist it is necessary to find out whether his control model is capable of realistically modelling regulating structures and the response of the water system to control actions
2. The current control strategy will always be the base-line control strategy against which new control strategies will be measured for their effectiveness and efficiency.

3. If considerable local (in time or place) differences are found between the modelled states and measured states, this may help to identify events and locations where or when operators decided to deviate from the normal control rules, for example, to anticipate extreme events

3.4.1 Input data based on end-use of model

However obvious, it is not common practice to use the same input variables and data sources for calibration and validation as will be used in operational tasks of the model, and indeed it is not trivial to realise in many cases.

A first limitation is often that multi-year time series data are available only from the (old) ground stations, while the model in the end will be fed with data from enhanced ground station networks or remote sensing data such as radar and satellite data.

The second limitation is that in operational forecasting applications the models are often forced with several sources of data for the same variable. For example precipitation input for rainfall-runoff modelling for flood forecasting could use ground station data up to $t = 0$, radar data up to $t = +2$ hrs, and quantitative precipitation forecast (QPF) data up to $t = +15$ days. Still, the model is usually only calibrated with the ground station data, while temporal and spatial scale differences may well influence the performance of these models with other data sources. If possible, the same source of data should be used for calibration, validation and application. If this is not possible, then a comparative analysis of the data should be made (Van Andel et al., 2009[a]), to assess whether different sources can be used for calibration, validation and application directly, or whether scaling of the data, or combined calibration and validation is necessary.

3.4.2 Framework for modelling controlled water systems

The framework discussed in this section concerns the model construction, calibration and validation phases of the modelling process (Abbott and Refsgaard 1996, p. 24) and emphasizes possibilities of iteration within these steps (Nash and Sutcliffe 1970). The modelling framework suggested for modelling of controlled water systems focuses on the problem of increased degree of freedom, because of the control structures. The solution depends on having more, and more reliable, measured data available (as is often the case for controlled systems). The measured data allows for two modelling steps that are not generally feasible and necessary with hydrological or hydrodynamic modelling of natural systems.

The first additional step is that with the increased availability of data, the model validation can be expanded from testing the model for the target variable for a period or event that was not used in the calibration, to a validation of the non-target variables for long term simulation periods. This extended validation allows identification and visualisation of any processes that might have been omitted or wrongly presented in the model. In other words, it allows visualisation, discussion and modelling of those processes in the water system that have been overlooked or are simply not known (Figure 3.7). In the proposed framework, it is suggested not to leave any of these deviations in the extended validation unresolved and un-modelled.

The second additional step can be taken if the increased data availability for a number of control structures clearly separates one sub-system from another. It is then often possible to replace sub-system models by time series data input, to enable model calibration of one sub-system at the time. This reduces considerably the danger of correcting one wrongly modelled sub-system or control structure by adjusting wrong parameter values to connected sub-systems or control structures as well.

Together with data acquisition, model set-up, calibration and validation, the modelling approach (Figure 3.7) is to first estimate all (physically based) parameters on the basis of the expert knowledge and available data, second, to calibrate the model, third, to compare the modelling results to check for trends that indicate that some processes have not yet been modelled (data driven approaches can be used), fourth, to model these deviations with either physically based (known and separable processes) or data driven (unknown or un-separable processes) model components, and finally to calibrate again the estimated parameters. When several sub-catchments have to be modelled, this methodology has to be used starting with the target variable (often at the downstream-end of the system), using available measured data of sub-systems as input, and then step by step to replace the measured input with models, because measured data or external predictions may not be available in operational forecasting mode.

In Section 4.4 the framework is applied to improve a water system control model of the Rijnland water system, in the Netherlands (Van Andel et al., 2009[a]).

3.5 Strategies for anticipatory water management

The available strategies for AWM can be described in three groups. These are discussed in the following sections.

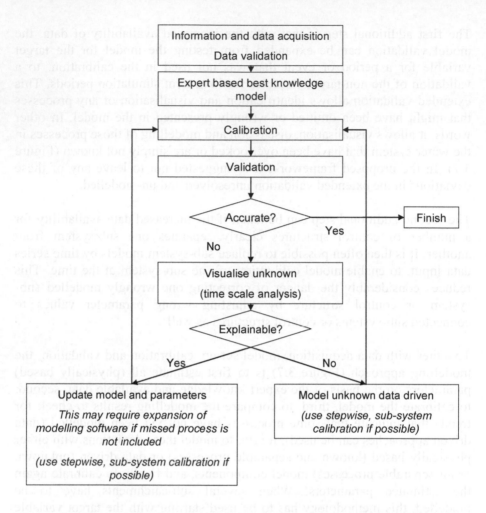

Figure 3.7 Framework for modelling controlled water systems. Visualisation, discussion and modelling of the unknown processes are key. If the processes can be identified (e.g. by error analysis for different time scales) and isolated after visualisation and discussion, they can be represented by an internal or external physically based model, if not, a data driven approach can be used.

3.5.1 Rule-based

Rule-based strategies consist of a combination of heuristic rules and pre-defined anticipatory actions, e.g. "if this than do that, else do something else". Examples could be decision trees when deciding whether to switch from normal management to anticipatory management (van Andel et al., 2008[a]), in combination with classes of pre-defined sub-optimal management strategies to minimise the damage of false alarms. In this respect, it is important to update the decision frequently.

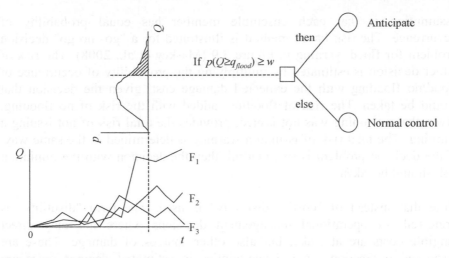

Figure 3.8 Fictitious example of a decision rule, based on three members of an ensemble hydro-meteorological forecast.

The decision tree accounts for probability (p, w) following Krzysztofowicz (2002) by choosing a threshold number of ensemble members (F_i) that forecast an alarm generating hydrological load or level (Q, $qflood$) within a certain lead-time (Figure 3.8). Instead of or in addition to this probability threshold type of rules, statistical measures (first moment, second moment, etc.) describing the estimated Pdf, can be used to relate the forecasts to pre-defined strategies.

3.5.2 Pre-processing of ensemble forecasts to deterministic forecast

In this approach the ensemble rainfall forecast is taken, and interpreted to determine the inflow volume that identifies the control action for the coming control time step. The most intuitive method is to take the average forecasted precipitation for every time step from the ensemble forecast. Deterministic optimisation methods can then be used for the strategy.

3.5.3 Risk-based

Risk based strategies refer to the use of decision rules on the basis of the estimated probability of occurrence times the estimated associated cost. The most widely used risk based decision rule is to decide on the alternative that has the minimum risk. The reasoning behind this decision rule is that if you apply it consistently over time, the actual cumulative total cost, after a long period, will be minimum as well. This will only be true if the probability of occurrence, e.g. the PDF of a flood event, can be forecasted accurately enough. For risk-based decision making with ensemble forecasts the assumption is that each ensemble member has equal probability of occurrence. The risk based method is illustrated for a "go- no go" decision problem for flood warning in Figure 3.9 (Maskey et al., 2008). The risk of either decision is estimated by multiplying the probability of occurrence of flood/no flooding with the expected damage cost, given the decision that would be taken. The risk of flooding, added with the risk of no flooding, given that a warning was not issued, provides the total risk of not issuing a warning. The total risk of issuing a warning is determined in the same way. If the decision problem is risk neutral, then the decision with the minimum risk should be taken.

Note that instead of "cost", "disutility" is used. The term "disutility" is preferred for operational management decisions where not only direct tangible costs are at stake, but also other sources of damage. These are discussed in Section 3.6.2. Uncertainties in estimated damage costs are generally high, but there is generally not much data or information about this uncertainty and methods for incorporating these uncertainties (as well as which uncertainty (not) to take into account) are being developed, debated or still need to be developed.

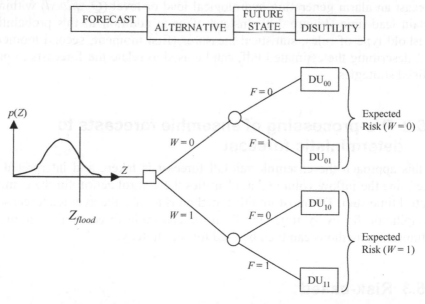

Figure 3.9 Risk based decision tree for flood warning, where the alternatives are W = {0, 1} and the future states of the system are F = {0, 1}. W = 0 and W = 1 imply "do not issue warning" and "issue warning", respectively. Similarly, F = 0 and F = 1 imply "the area is flooded" and "the area is not flooded", respectively. (Cited from Maskey et al., 2008)

3.6   Cost-benefit of selected AWM strategies

3.6.1 Dynamic cost-benefit analysis

The next stage of the screening is an evaluation of a number of suitable AWM strategies. Although the forecast verification informs us about the numbers of hits, misses and false alarms, it is not known how much is gained from a hit and how much is lost from a false alarm. Therefore, the verification analysis is suitable for comparing different forecasting systems and decision rules, while it does not determine whether it is beneficial for a water authority to implement AWM. To resolve this an evaluation of the hits, missed events and false alarms is needed. In meteorology this is often done with a "cost-loss" analysis where the results are shown for a range of cost/loss ratios. In such an analysis the "cost" refers to the costs of anticipatory management actions (such as an evacuation) and "loss" refers to the damage costs of a critical events when no measures are taken. Although working with cost-loss ratios is good for inter-comparison, it is not suitable for deciding whether or not to adopt a new strategy, because the method does not work with absolute evaluation.

Cost-loss ratios can be based on absolute damage estimates, but the assumption of a fixed ratio is too much of a simplification for AWM. In water resources management every event is different and so are the AWM actions. Therefore there is a need to prepare a dynamic, absolute valuation of the operational water management. This would estimate the total value (cost) over a given analysis period. Then current and alternative operational management strategies can be compared in a cost-benefit analysis (Figure 3.10).

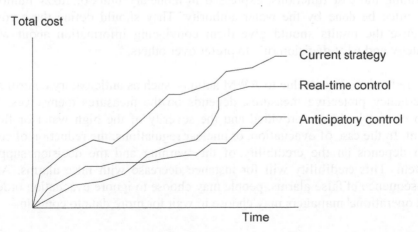

Figure 3.10 Total cost estimation for alternative operational water management strategies

3.6.2 Sources of damage

There are wide varieties of costs and of ways for these to be included in a cost model. Traditional damage functions of direct tangible costs (e.g. damage to houses because of flooding) are mostly used. In addition more and more indirect tangible (disruption of economic activities) and intangible (loss of life, social disruption, loss of credibility of a warning system when false alarms are being issued) damage costs are also taken into account. The tangible costs are usually expressed mathematically as functions of water system state variables such as water level, whereas the intangible costs are often not expressed or expressed in terms of fuzzy membership functions.

Aside from the events, there are also continuous costs, such as operational costs. Operational costs can be related to power costs for operating structures, maintenance costs (lifecycle), environmental costs and social costs.

In this research we propose an extension of the traditional damage cost functions to the time duration cost functions. Mostly the damage functions are constant for every time step, while in reality damage can grow exponentially when the water system stays in an undesired state. For example, flooding damage to crops can be highly dependent on the duration of the inundation.

First, for the screening of new forecasting products and AWM strategies, the most prominent sources of damage that can be estimated directly with the same continuous modelling system as is used for the warning verification. Defining the cost functions, expressed in monetary units or fuzzy numbers etc, must be done by the water authority. They should define these costs, because the results should give them convincing information about what strategy and what decision rule to prefer over others.

The reduction of costs due to AWM actions, such as anticipatory control and emergency protective measures, depends on the measures themselves, but also on the lead-time provided and the severity of the high water or flood event. In the case of evacuation, or unclear regulations, the reduction of costs also depends on the credibility of the warning and the decision support system. This credibility will for instance decrease with false alarms. As a consequence of false alarms, people may choose to ignore evacuation orders, and operational managers may choose to wait for more data to come in.

3.6.3 Anticipatory Water Management modelling

To perform the cost-benefit analysis, the continuous simulations with the water system model have to include the emulation of the control strategy as well. A continuous simulation of operational management of the water system has to be made. In the case of an AWM strategy, this means that meteorological forecasts have to be input to the hydrological/hydrodynamic model and the water system control model with the normal control rules. Decision rules when the resulting warnings produce a temporary shift to AWM control rules should be incorporated. Then the AWM control actions are modelled and the effect on the water system state variables will be known. This results in a continuous time series of water system state variables, on the basis of which the total cost over the analysis period can be calculated.

In real-life application every time a new precipitation ensemble forecast comes in, the ensemble water level forecasts will be updated (or even more often, e.g. for every time that the control strategy needs to be updated (e.g. one hour)). If a yes/no decision has to be made first, whether AWM or normal control will be applied, these updated EPS water level forecasts will be done assuming normal control to see whether the water levels remain within the target range. Updating can be important to reduce the number and duration of false alarms. Note that because normal control is applied for the forecast, the modelled system state will immediately start returning to its normal range, regardless of current anticipatory actions. As a result, for fast responding systems, and far forecast horizons, the effect of updating forecasts will be limited.

On the basis of the warning verification and with the help of global optimisation methods a limited number of suitable decision rules should be evaluated in this way. The resulting total cost estimates can be compared with the costs of the current control strategy and other non-AWM strategies. This overview of costs of a number of suitable AWM strategies gives valuable information to the water authorities about it may cost to adopt AWM and what decision rules are efficient and what are not.

This ends the screening of AWM. If the screening results are satisfactory, a further optimisation of the control strategy and decision rules needs to be done at the next stage.

3.7 Optimisation of Anticipatory Water Management

The main problem of optimising the control strategy of water systems is that the variation in potential control strategies is usually very large and dependent on a large number of different decision variables, not only on the wide range of an individual variable. This multi-year optimisation problem, in which per day several ensemble predictions are available and the best management strategy for the entire period needs to be defined, cannot be captured in an analytical optimisation model. Therefore, global optimisation methods with smart search methods, like evolutionary approaches, are used.

3.7.1 Objectives

For each case study the objectives need to be defined and agreed upon before optimisation can take place. Usually the optimisation problem will be a multiple-objective problem. In the case of flood control of a reservoir, example objectives could be to:
1. Minimise flood damage cost
2. Minimise total damage cost

For a multi-objective problem (with conflicting objectives) to result in one optimal solution, weights have to be given to the objectives. Because this is always a heavily debated (before and after) process, of which the consequences for the end result are not known up-front, preference is often given to the provision of multiple possible optimal solutions in a Pareto Front (Coello, 2005; Barreto et al., 2006).

3.7.2 Parameterisation of AWM strategies

The parameters that have to be optimised are the variables that make up the warning and operation rules for the AWM strategy. Therefore the strategies discussed in Section 3.5 have to be parameterised. For example, for a strategy following a simple ensemble based threshold decision rule for the early lowering of a storage basin water level to a fixed level:

If forecasted probability P(water level Y days from now $> H$ m+Ref) $> N$, then at A days from now start lowering water level to H_a m+Ref.

the decision parameters become:
-   Water level threshold (H)
-   Forecast horizon(Y)
-   Probability threshold (N)
-   Anticipation time (A)
-   Anticipation water level (H_a)

However, when elaborating this strategy, an update frequency (control decision time step) and a rule on how to deal with inconsistency in the forecasts are needed. It can be seen that for complete control strategies the number of optimisation parameters grows fast. Therefore, it has to be considered whether sub-sets of the parameters can be optimised separately.

3.7.3 Optimisation using perfect forecasts

In some cases of sub-optimisation, some of the decision variables can be selected on the basis of perfect forecasts. In this way, the interpretation of the probabilistic precipitation or water level forecasts is separated from the actual anticipation action, since the decision to take action has been taken. The perfect control actions for a given system load do not depend on the forecast, but are in fact system characteristics. Once the decision variables that are system characteristics have been defined, these can be separately optimised, using the water system control model with perfect (measured) forecasts. Examples of these parameters are the optimal anticipation time and the update frequency. In addition optimisation with perfect forecaststs shows the maximum benefit that can be achieved by applying AWM.

For example, in the case of flood control, the maximum anticipation time (control horizon) required is determined by the maximum flood event in the verification analysis and its antecedent and post event conditions. These together determine how much time is needed to pump out the excess volume of water (before (and or) after the event). Further expanding the control horizon beyond this time has no effect on the analysis. Therefore the desired maximum control horizon should be determined with the design storms. Design storms do not necessarily have to be part of the verification archive. Storms can be defined for larger return periods such that they take into account expected climate change. Note that reference is made here to the control horizon assuming a perfect forecasting system. A larger control horizon may be chosen in reality, when working with imperfect forecasts, to account for the possibility that the event is forecasted too late.

3.7.4 Optimisation with actual forecasts

A wide range of strategies can and may have to be tried in order to come up with a reliable optimal control strategy for real, imperfect, forecasts. The different kinds of strategy as discussed in Section 3.5 have to be optimised separately. Rule based AWM strategies can be optimised by evolutionary search methods, such as Genetic Algorithms. Rule based AWM strategies need optimisation of two main components. The first is the interpretation (pre-processing) of the hydro-meteorological forecasts in general (long-term strategy optimisation). The second concerns the short-term optimal control actions with the regulating structures. Through the simulation of a particular strategy for a long historic period for which measured data is available, the objective functions can be estimated. A Pareto front of the multi-objectives can be produced (Figure 3.11). In this way the short term, real-time, management actions, and the long term operational strategy can be optimised simultaneously (layered optimisation).

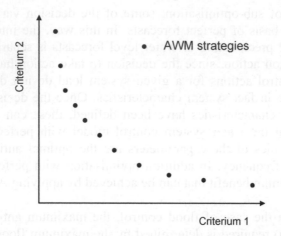

Figure 3.11 Pareto front for a 2-objective (criteria) optimisation problem with AWM strategies

3.8 Decision making for policy adoption of AWM

3.8.1 What-if analysis

The analyses described will be informative and convincing because they make use of measured data and archived forecasts. Therefore, they clearly show what could have been done with AWM in the past to improve the management of critical events. However, for policy decisions to adopt AWM we need to consider what these results in the past tell us about the future. Design storms could be used, but the associated forecasts are not available, so there appears to be no guarantee that a developed AWM strategy will perform just as well for more extreme events, in changed climate conditions. However, the atmospheric models are physically based and make use of real-time data assimilation. This implies that if these types of events occur in places they did not occur before due to climate change, they will be forecasted by the models just as well. Therefore, it can be assumed that the performance of AWM, achieved with archived data will not deteriorate in the future. With the continuous further development of the hydro-meteorological numerical modelling, it may even be hoped that performance will only improve.

What is more important, however, is that the reliability of the expectations we will get from these analysis increases with increasing simulation periods. Therefore archiving of measured data and re-analysis and hindcasting are becoming ever more important.

3.8.2 Re-analysis era

Developments in water system modelling have resulted in reduced computational demand, and at the same time developments in parallel and grid computing, together with the ongoing increase of processor speeds have reduced computational time. Together these developments greatly enhance the use of simulation models, scenario analysis, and optimisation in both operational and strategic water management. The analyses described in this work show how advantage can be taken of these new opportunities in practical additions to the current analyses of the water authorities.

3.9 Framework for developing Anticipatory Water Management

The methods described in the previous sections in response to the knowledge gaps described in Section 2.6 together form a framework to develop, evaluate and adopt an Anticipatory Water Management strategy for a given water system (Figure 3.12).

This framework supports water managers to evaluate a new forecasting product for application in Anticipatory Water Management (AWM). The main part of the framework consists of steps that perform a screening of new forecast products and control strategies. The outcome of the screening should be twofold: It should indicate the range of suitable decision rules for AWM and it should benchmark the proposed operational management strategy against current management and alternative strategies.

Then if the forecasting product is selected and suitable AWM strategies seem to be available, as a second stage, optimisation of the AWM strategy can be performed.

After this follow the stages of implementing the decision support system, learning how to use it, and building the confidence of the operational water managers. At some stage the new operational strategies need to be incorporated in the legislative policies of the particular water authority.

In general technological advances bring shifts of responsibilities from one group to another. This poses strains on an organisation that adopts technological change, as with any other change. These strains have to be dealt with by involvement of all affected groups from the beginning, building consensus and trust that the changes are for the better, and that the new tasks for the different groups are clearly defined and satisfactory to all.

In chapters 4 and 5 the AWM framework is tested in case studies.

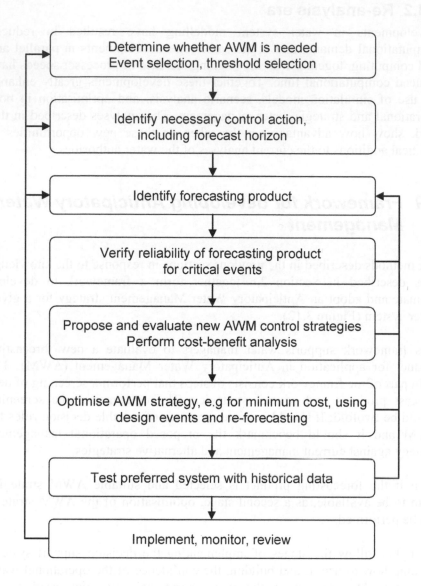

Figure 3.12 Framework for developing Anticipatory Water Management. The main part of the framework consists of steps for screening of new forecast products and control strategies. If new control strategies perform well, in the next step the optimisation of the AWM strategy can be performed.

4 Case study 1 - Rijnland Water System

4.1 Introduction

Rijnland is a polder area in the western part of the Netherlands, bordering the North Sea (Figure 4.1). The total area is about 1000 km^2 of which 72% is occupied by low-lying land-reclamation areas, 15% by free draining areas and 8% by dunes. A storage basin consisting of inner connected canals and lakes, occupies 45 km^2. The storage basin serves to collect all the excess water of the Rijnland area, before it is discharged to the main water system of the Netherlands and finally to the North Sea. The low-lying areas would be subject to flooding if they were not protected by dikes and the excess water not pumped to the storage basin. The water level in the storage basin is kept between predefined bounds, mainly by the daily operation of four large pumping stations: Halfweg, Katwijk, Spaarndam and Gouda. The total capacity of these four pumping stations is 154 m^3/s (13.3 mm/day). In case of extreme events also pumping station Leidschendam may be used to discharge water to the Delfland Water Board (8 m^3/s).

The area consists of urban and rural parts. The rural parts can be sub-divided into areas committed to horticulture, agriculture, and grass lands. The dominant soil types are sandy in the free-draining and dune areas, clay in part of the land reclamation areas, and peat in the main part of the land reclamation areas.

Excess water is discharged from the urban areas to the main storage basin through waste water treatment plants. Combined sewer overflow discharges end up in the drainage network of the rural areas, and through small pumping stations are pumped to the main storage basin.

During and after rainfall events, excess water from the land reclamation areas is pumped directly to the main storage basin by over 200 small pumping stations. Further hydrological load to the storage basin comes from the free draining areas and excess water from the neighbouring water board "Woerden", which is discharged to the Rijnland storage basin through an inlet (Bodegraven inlet).

During summer and dry spells, the channelled storage basin is flushed by combined operation of smaller inlets and sluices, mainly Gouda inlet and KvL sluice, and the four main pumping stations.

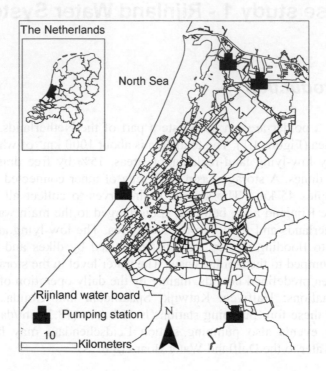

Figure 4.1 Principal Water-board of Rijnland: controlling a low-lying regional water system in the western part of the Netherlands. A channelled storage basin collects all the excess water of the area. The water level in the storage basin is controlled by four pumping stations.

The response time, taken as time from peak of precipitation to the end of the first half of the associated pumping period (representing the peak of the run-off response), varies between 0.5 to 1.5 day. Time to empty, taken as the time from the peak precipitation to the moment that the water level is back to normal, is about 3 to 4 days for big events (60 mm/3days).

A simulation model of the water system is used in a decision support system (DSS) for operation of the four pumping stations to keep the storage basin water level within a 0.05 m target range (Figure 4.2).

The following sections describe the results of each step in the framework for developing Anticipatory Water Management (Figure 3.12 as applied to the Rijnland water system.

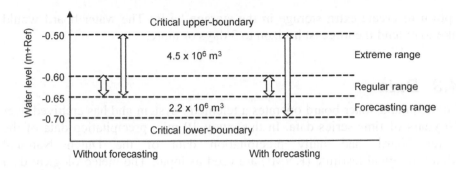

Figure 4.2 Water level control of the Rijnland storage basin, with and without forecasting. When using forecasts and temporarily allowing lower water levels, extra storage of 2.2 x 106 m3 can be created before the extreme event occurs.

4.2 Problem description

The Rijnland area has faced both extreme precipitation events (1998, 2000) and droughts (2003). Research has concluded that the required safety level against floods is no longer being met. The estimated probability of exceeding the critical water level in a year is more than 0.01, and therefore remedial structural measures are planned by the Principal Water-board of Rijnland (Rijnland, 2000). Emergency storage basins are to be allocated and the pumping capacity is to be increased by 40 m^3/s (3.5 mm/day), which is 26% of the present pumping capacity (13.3 mm/day). In addition, the Rijnland water board would like to optimize the operational flood control of its water system. The proposed anticipatory measure is to create extra storage in the basin when extreme hydrological loads are expected. Anticipatory pumping can lower the storage basin water level below the regular range (Figure 4.2) before the extreme event occurs. The pumping would create extra storage, which is comparable to that of the planned emergency basins. Swinkels (2004) used offline control simulations to show that for the extreme event of 2000, during which the −0.50 m + Ref (Dutch reference level ~ mean sea level) level was exceeded, a forecast horizon of at least 3 days would be necessary with an allowance of 0.08 m extra storage, to prevent the water level from exceeding the maximum permitted value. The problem is that low water levels may have adverse effects, such as hindrance of navigation and damage to houseboats, and for very low water levels the problem is the risk of embankments becoming unstable and soil subsidence, bringing damage to nearby houses (generating an economic risk). The water board started to apply heuristic rules for anticipatory pumping to increase safety against floods. A daily precipitation threshold of 15 mm was chosen as the alert threshold, using the precautionary principle that the forecast may underestimate the precipitation. If the 1-day precipitation forecast exceeds this threshold, early pumping is considered an option to create extra storage in the storage basin. The water board would like to extend the forecast horizon to 3 days or more.

4.3 Data

The Rijnland water board operates a telemetry system and has archived over 20 years of time series data. In this study 10 min. precipitation data of the water board, and daily precipitation data of the Dutch National Meteorological Institute (KNMI) are used as input. The meteorological data is processed with HydroNet DSS (HydroNet, 2009) The daily data is validated by the KNMI and has no missing data. The 10 min. precipitation data contains outliers and missing data. When these occur, only the daily data of the KNMI are used. For evaporation data the daily Makking reference evaporation from 3 KNMI ground stations is used. These data are validated and contain no outliers. If a station contains missing data, it is not considered. For the analysis period of this study there was always at least one of the three stations functioning properly. For sub-system calibration (Section 3.4.2) also daily flow data from the water board is used. This data is validated and contains no missing data or outliers.

Rainfall

In the future operational DSS, rainfall radar may be used as input to the water system model. However, for the calibration and validation of the model up to t = 0 the available archive of radar data at the time of calibration was limited to one year, while the available ground station data was (more than) 7.5 years. With respect to rainfall input the model is fully lumped meaning that the area average rainfall is used as input. This is because the area average water level is the target variable for operation requirements, and because in the storage basin of connected canals, local effects of rainfall flatten out quite fast. To check whether the use of radar would not change the model performance that is calibrated with ground station data, the area average rainfall from both sources is compared (Figure 4.3).

In the forecast verification analysis, daily precipitation data for 16 ground stations from the KNMI and 10 min precipitation data from 6 stations from Rijnland is used to estimate hourly area-average precipitation in the Rijnland area. The Thiessen average daily sums of the KNMI stations are preserved. The hourly distributions of the daily sums are taken from the hourly Thiessen average of the 10 min data.

The radar data is stored in 3 hourly sums, updated every hour, for a grid of regular 2.5 km by 2.5 km cells. For both sources the data has been aggregated to daily sums for December 2004 and presented in Figure 4.3. The result shows little difference between the sources (3-hourly rainfall data is provided by the KNMI after calibration with ground station, so this is not very surprising). The maximum daily difference is 0.9 mm, the difference in the month sums is 0.6 mm. Therefore in this case study the archive of ground station data can be used for the calibration and validation, without expecting too many problems when switching to using radar data in the operational application. Hourly precipitation data is used.

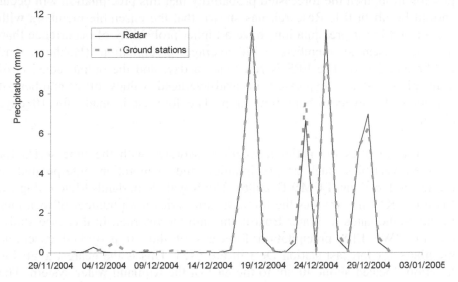

Figure 4.3 Comparison of radar and ground station precipitation estimates for the Rijnland area. The graphs show close resemblance for both dry and wet periods.

Evaporation
The input evaporation data is provided as daily Makkink Reference Evaporation from three KNMI meteorological stations in the area. Forecasts of evaporation are not used, because of the limited effect on peak discharge events within a 10-day horizon.

Precipitation forecasts: ECMWF EPS
Precipitation forecasts of the ECMWF EPS are used. The ECMWF began producing EPS forecasts operationally in December 1992 with 33 members (different runs) of their global circulation model (Molteni et al. 1996). The EPS is under continuous development. Since 1996 the model has been run 52 times for each forecast: one run with a high spatial resolution (operational run), one run with the EPS spatial resolution and un-perturbed initial conditions (control run), and 50 ensemble members with perturbed initial conditions. In 2000, the spatial resolution of the operational run increased from roughly 60 km to 40 km and the EPS from 120 to 80 km. In 2006, the spatial resolution of the EPS was further increased to 50 by 50 km. Since 1998, also a scheme for model error has been included (stochastic forcing).

The development of the operational EPS is described on the ECMWF website (ECMWF, 2007).

The assumption is that the perturbed initial conditions are determined in such a way that the 50 ensemble members are equally likely to occur (Persson and Grazzini, 2007). If, for instance, five ensemble members predict a certain precipitation, then the forecasted probability that this precipitation will occur should be about 0.1. Research has shown that the ensemble members with higher and lower precipitations have a higher probability of occurrence than do the ensemble members with average precipitation (Bokhorst and Lobbrecht, 2005). The EPS is run twice a day, and the output consists of atmospheric states, expressed in grid-averaged values of a number of variables, for every 6-hour time step. The forecast is made for 10 days ahead.

ECMWF supplies the national weather institutes with the time series for selected variables, such as precipitation and evaporation, interpolated to requested locations (ECMWF, 2006). The Royal Netherlands Meteorological Institute (KNMI) provides the forecast time series for a number of locations in the Netherlands to Water Boards through the Internet. In this case study, the ECMWF EPS precipitation forecasts of the 50 perturbed ensemble members for forecast station De Bilt are used. This is the nearest available location, located about 40 km to the east of the Rijnland water system. The forecasts are compared with area-average measured precipitation to avoid large spatial scale differences (local extremes versus large-scale precipitation).

Downscaling techniques and bias analysis have not been applied in this research; the aim is to establish what can be done by adjusting the decision rules with the ECMWF EPS forecasts as they are.

4.4 Water system control model

4.4.1 Model structure

The Rijnland water system is modelled with the Aquarius modelling software (van Andel, 2009[a]). Aquarius is an object oriented non-linear reservoir model. Surface water, layered soil columns, ur-ban areas, and green houses make up the elements of the reservoirs. Most important additions to the rainfall-runoff process are the control structures. Weirs, sluices, inlets and pumping stations can be modelled, including the control methods used, such as PID, local switch on/off levels, and global control (Lobbrecht, 1997).

The Rijnland water system is modelled with three distinct sub-areas, representing the storage basin and dunes (Rijnland storage basin), land-reclamation areas with clay soils (Rijnland Polder 1), and land-reclamation areas with peat polders (Rijnland Polder 2). During events with excessive rainfall, Rijnland receives excess water from the neighbouring water board (Woerden). The Woerden area is modelled similarly with a storage basin area, and a land reclamation area (Figure 4.4).

The characteristics of the sub-systems and the control structures have been provided and estimated by the Rijnland Water Board.

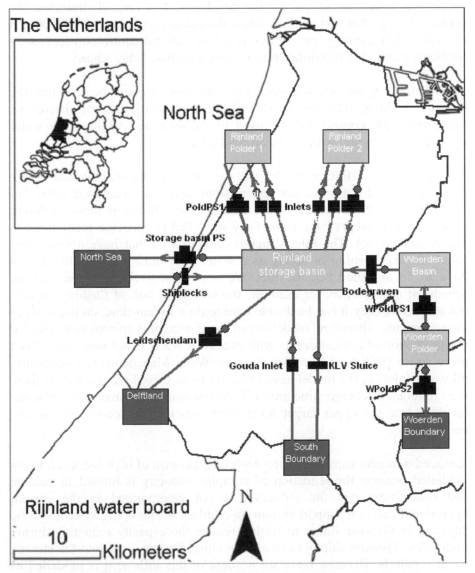

Figure 4.4 Aquarius water system control model of Rijnland (Yufeng, 2003)

4.4.2 Control strategy

The envisaged application of the Aquarius model is to operational use in decision support for the operation of the four main pumping stations, and for research to compare control strategies. The latter is done on the basis of long term simulation of the Rijnland water system and its operation for a multi-year period. Therefore the simulation period applied for this building this model is 1997 - 2002 for calibration and 2003-2004 for validation, for which a comprehensive archive of water system data and reports was available.

To evaluate the model's capability to be used as a decision support tool for operational management and for the development of control strategies, the control strategy that was applied during the simulation period (1997-2004) is modelled. This strategy can be described as a global control strategy, with qualitative inclusion of rainfall-runoff forecasts up to 1-day a head.

When analysing the water system data, however, it could be seen that the strategy could be well simulated with local automatic control. Therefore, local automatic control will be used to calibrate and validate the water system control model with the 1997-2004 data.

The application of forecasting for one day or more, requires that the inflow into the storage basin is not taken from real-time measured data, but modelled as well. Therefore, structures that have a discharge function during excess rainfall events have to be fully modelled (to provide predictions in operational mode). The inlet structures and sluices that have a regulatory function, like flushing or acting as a water inlet during dry spells, do not have to be modelled because they are not responsively operated, but are scheduled tasks. When, for example, the scheduled task of flushing appears not to be necessary it can be decided not to do it in real-time, on the basis of measurements. Therefore, modelling of these structures control strategies for the prediction of critical events with excessive water is not necessary. Even more so, to predict whether Anticipatory Water Management is necessary, all water inlets in the model have to be set to be closed (zero inlet). If then, the modelled discharge structures still do not manage to maintain the water levels below the upper target level, then anticipatory control actions are needed.

Reduced pumping capacity during high tide, because of high sea level, is not modelled because the reduction of pumping capacity is limited in volume and time (tide) and the reduction is not incorporated in the current operational decision support system. In addition the pumping system will be adjusted in the near future to further reduce the capacity reduction during high tides. Through shiplocks small amounts of water come into the storage basin regularly. Because there are no data of this inflow, it is modelled as part of an constant external inflow to the Rijnland storage basin, estimated by the water board on the basis of information on hydrological loads and yearly water balance studies.

4.4.3 Model calibration

In the Rijnland case study the target variable is water level in the storage basin. For calibration of the water level data a critical rainfall event with high storage basin levels, in November 2000 was selected.

Local automatic control is modelled, therefore, the control parameters to be adjusted are the on- and -off set points of pumps and inlets, and open-close set points for sluices. Whereas the set points of control structures can be known or inferred from the water board's information on the applied control strategy, the soil in- and outflow resistances are highly uncertain. Soil type gives only ranges of possible resistances, and on top of this, soils in most catchments are highly heterogeneous, often further complicated by a varying drain network, which makes the area averaged soil in- and outflow resistances highly uncertain. Therefore, the latter are the main calibration parameters. For the three Rijnland sub-systems, two land reclamation areas and the storage basin, three soil layers have been defined for each of which separate in- and outflow resistances are defined. This makes a total of 18 calibration parameters. The Woerden sub-system has one reclamation area with three soil layers, which makes six parameters to be calibrated. In addition, the switch -on and -off levels of the Rijnland storage basin pumping station that were received from the water board were further refined through calibration (12 parameters).

Water level peaks are the most important events to be calibrated, because the main operational purpose of the Aquarius Rijnland model will be decision support for flood control. The modelled and measured slope of the rising water level have to match well, the peak water level has to be modelled accurately, and the slope of the receding water level is also a good indication whether the runoff process are modelled well. The normal flow periods are important indicators for the quality of the model as well. The slope of the rising water level after discharge control actions have stopped, and the slope of the receding water level when discharge is taking place are determined by the rainfall runoff process, soil moisture content and groundwater level, and groundwater in- and outflow resistances. Therefore the criteria for calibration of the Rijnland model are visual similarity of measured and modelled water level peaks, and normal flow periods.

Figure 4.5 shows the calibration result for the event of November 2000. The rise of the water level around 10 November has been modelled especially well. Note that the deviation in the beginning of the peak can be the result of manual operation in reality, which defers from automatic local operation in the model. After the peak the modelled water level recedes not as fast as the measured water level. This can be the result of errors in the rainfall input, but also it could be the result of sub-optimal ground water flow resistance parameters, initial conditions, or even surface area errors. As is often the case in both hydrological and water system control modelling, if there are still errors remaining, after the best of knowledge has been put into the model, it is unknown what is causing the remaining error.

The simulation period for calibration has to be long enough to prevent additional errors due to wrong initial state before the event. Note that when changing the simulation start time, it needs to be checked that the initial conditions (water level, groundwater level, soil moisture content) match the new start date.

Low flow periods, such as in February 2000 (Figure 4.6), were simulated accurately as well. Slopes of rising and receding water levels match well. Again, small differences remain, but these most likely come from the difference between manual and full automatic control.

As a next step validation was performed.

4.4.4 Model validation

The model has been validation for the year 2002 on the basis of monthly discharge volume through the four main pumping stations. For comparison the annual report monthly values have been used (Figure 4.7). April to August, summer months, are the relatively dry months in the Netherlands. Although the water level modelling results were reasonable the monthly pumped discharge volumes are clearly under-estimated by the model. In October and November 2002 the pumped volume is over-estimated by the model. For the other months the validation is satisfactory.

In some cases, for the water boards, this modelling result could be considered satisfactory, because in general flood events occur less frequent in summer months than in winter months, and because despite the poor validation in October and November the modelled water levels compared quite well. Note that this is evidence of the point made in the introduction that water-system control modelling has a high degree of freedom.

On the other hand, Figure 4.7 could raise the question why the pumped volume is strongly underestimated in April and May and then gradually improves to result in an over-estimation in October and November, and whether this is a systematic error or not.

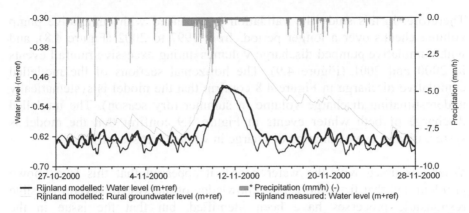

Figure 4.5 Calibration of Aquarius water system control model of Rijnland, the Netherlands, for a peak water level event in November 2000.

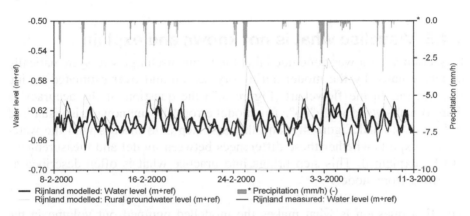

Figure 4.6 Calibration of Aquarius water system control model of Rijnland, the Netherlands, for a normal flow period in February and March 2000.

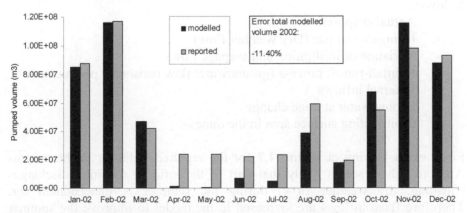

Figure 4.7 Validation of the Aquarius water system control model of Rijnland, on the basis of monthly pumped discharge volumes.

Therefore, in this study the validation is expanded with cumulative pump volume checks over a longer period, from 1997 to 2002 (Figure 4.8), and with cumulative pumped discharge volumes during excessive rainfall events in 2000 and 2001 (Figure 4.9). The horizontal sections of the modelled cumulative discharge in Figure 4.8 confirm that the model is systematically underestimating discharge volume in summer (dry season). The modelled discharge of both winter events in Figure 4.9 confirm that the model is systematically overestimating discharge in winter (wet season).

When discussed with the water board, it appeared that this is a known problem and that from the expert knowledge of the system several possible responsible processes have been identified, but that the issue in the modelling had so far not been resolved. In the following sections, following the approach proposed in Figure 3.7, the model results are further analysed to visualise the unknown prosesses and try to explain them.

4.4.5 Visualise what is not known and explain

In Section 4.4.4 it was concluded that the pump discharge was systematically underestimated by the model during dry season and over-estimated during low season. In the flowchart (Figure 3.7) the question of the accuracy is therefore answered with "no". The model error, "the unknown", is visualised in Figure 4.8 and Figure 4.9, and the next step is to discuss with the water system experts whether these differences between model and measurements can be explained. This step brings into practice what is often described as "learning from models".

The first question is what makes the modelled pumped out volume in the summer too low. The answer requires the uncertain input variables or calibration parameters that can cause errors in the long term (monthly, seasonal) processes to be identified. The variables and parameters are as follows:
- Actual evaporation
- Human water use (Dry weather flow)
- Variation of infiltration and seepage flows
- Rainfall-runoff process (groundwater flow resistance parameters)
- External inflows
- Groundwater storage change
- Contributing surface area in the dunes

First, when looking at Figure 4.7 for the reported volumes in the months April to July it seems likely that part of the inflows and inlet discharges (flushing, dry weather flow) during the summer is unaccounted for. Therefore fixed inflows are increased in the model to improve the summer pumping. However, because these fixed flows are assumed constant for the

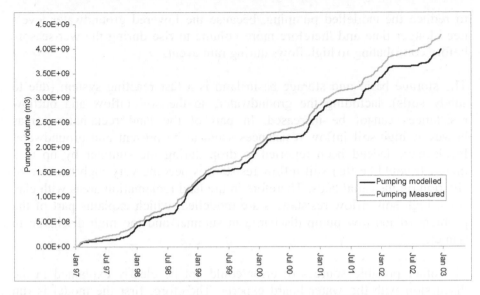

Figure 4.8 Cumulative pump discharge volume from the Rijnland storage basin. Modelled volume is too low, because of underestimation during the dry summer seasons.

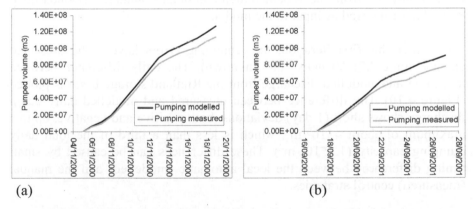

(a)   (b)

Figure 4.9 Cumulative pump discharge volume for events in the wet winter season in 2000(a) and 2001(b). For both events modelled volume is higher than measured volume, indicating over-estimation of the model during excess water events in the wet season.

whole year, this exacerbates the problem of too much pumping in the wet season.

Secondly, increasing the soil inflow resistance helps to increase pumped volumes in the dry season, because it reduces flow from the surface water to the ground water. Groundwater is allowed to drop a little during dry season, and surface water levels are maintained more easily. Increasing the soil inflow resistance in the early winter months (September, October) also helps to reduce the modelled pumping, because the lowered groundwater levels need longer time and therefore more volume to rise during the wet season, before contributing to high flows during rain events.

The storage basin and storage basin land is a fast reacting system (due to sandy soils), including the groundwater, so the soil inflow and outflow resistances cannot be increased. In part of the land reclamation areas however, high soil inflow resistances seem to be present and groundwater levels have indeed been reported to drop during the summer by up to a meter. It could be that soil inflow resistances become very high because of siltation on the canal beds. Therefore in the land reclamation areas with clay soils, high soil inflow resistances are modelled, which explains part of the problem of too low pump discharge in summer and too high discharge in winter.

The other possible sources of error could not be clearly explained in the discussion with the water board experts. Therefore, first the model is run again with the adjustments mentioned, following the left track of the modelling framework (Figure 3.7). To allow the modelling of the Rijnland area separately (Figure 3.7, sub-system calibration), the available measured in- and outflows from the Woerden area and other boundaries (Gouda inlet, KvL sluice) are used as input to the model.

Then, after this first iteration, the remaining errors have to be presented again (Figure 3.7: Visualise the unknown). The daily difference between measured and modelled discharge from the Rijnland storage basin is plotted (Figure 4.10). The difference between measured and modelled daily pump discharge shows short (1 day) deviations, both positive and negative, up to a maximum of about $4*10^6$ m^3, which is less than a third of the total daily pumping capacity ($1.3*10^7$ m^3). These differences can be caused by small timing differences between the local automatic (modelled) and the manual (measured) control strategies.

The effort is to find out the longer term processes (e.g. seepage, water consumption). These processes are expected not to vary too much from day to day, but do have a monthly change and variability over the year, gradually changing along patterns following the dry and wet seasons. One assumption often made is to discretisise between summer and winter season, because indeed in the Netherlands water management is changed from one day to the other when it is decided to switch from winter to summer target levels and vice versa. Still the natural process, cropping seasons, and domestic and industrial water uses follow more graduate trends that are often not known and not modelled in the water system control models for the water board.

Therefore a time scale analysis of the model errors has to be performed to filter out the fast processes that are to be calibrated later, and to capture the unknown slow varying (averaged) processes. When the time scale is scaled up to 10, 20 up to 90 days a clear sine pattern emerges (Figure 4.10). This sine pattern has a 1-year period, which was checked to hold for 6 years of data.

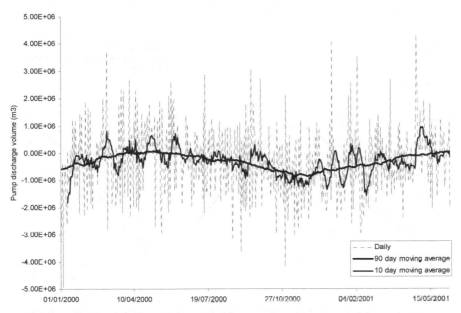

Figure 4.10 Time scale analysis of difference between measured and modelled pump discharge. At 90-days moving average a clear sine function with a yearly period becomes visible.

4.4.6 Modelling the unknown phenomena

Experts from the water board had two main notes on the remaining sine function error. The first is that the errors probably come from not fully capturing the soil and groundwater processes with the model. Small errors are inevitable there, and because of the large volumes involved these small errors cause big water level and volume balance differences between modelled and measured values. Secondly the sine could result from gradually changing groundwater levels in the dunes (including semi-controlled drinking water production) causing dynamics in the seepage flow to the polders. Unknowns in the actual evaporation throughout the year (dynamic land use, harvest times etc) and domestic water use (industrial water use is limited) may be additional causes of the sine error shape, although part of the changes in actual evaporation are covered by the monthly Makking evaporation crop factors in the model.

Since the model was already updated on the basis of the validation, and since the discussion of the sine function with the Rijnland water board representative did not result in a uniquely defined process that is causing the sine, it was decided to apply a data driven model for the sine function and use it to define the external inflows to the model.

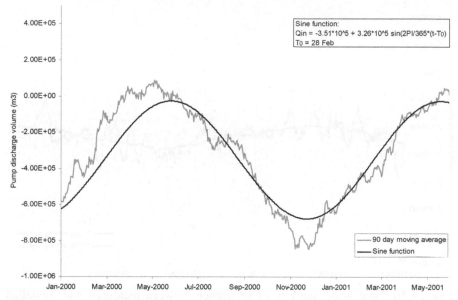

Figure 4.11 Sine function to model the slow processes error (90-days moving average) of the Rijnland Aquarius model.

The model is deduced for the Rijnland land reclamation areas and storage basin. The 3-parameter sine-function was optimised (Figure 4.11). Note that only part of the 6-year period is presented. The slow cyclic behaviour can be modelled well with the sine function. On top of this, there will still be other, short-term errors that will partly be compensated by calibration and are partly inevitable because of inconsistencies in the measured time series due to manual operation of the control structures. The sine function is added in the model as an external ground water inflow to the land reclamation areas. In the same way, the sine inflow function to the neighbouring area, Woerden, was prepared. A special feature there, is that there seems to have been a system or policy change in January 2001, after which much more water was discharged to Rijnland. This means that in the model two sets of control rules for the control structures are used (before and after 2001) and that two separate sine functions had to be prepared.

4.4.7 Final model results

After the physically based, and data driven adjustments were implemented the final steps are to prepare the model in the form it will be used operationally and to re-calibrate and validate this final model (Figure 3.7, second loop).

For the sub-system modelling, control structure models were replaced by measurements, but in the operational system not all these measurements can be used. The model is planned to be used for decision support in operational management by providing warnings for high inflow, and high water level events. Therefore, structures that have a discharge function during high flows have to be modelled. Structures that have a regulatory function during normal or low-flow periods are operated according to schedules, so these do not have to be modelled.

The Kocksluice or KvL sluice has no discharge function during excess hydrological loads to the Rijnland storage basin, so it is modelled using the measurement as input for times up to time 0. In predictive mode, the discharge through the sluice will be put to zero.

The Leidschendam pumping station is used regularly in summer with small volumes, for flushing of Delfland. In winter only it is used incidentally with peak events like the one in 2000. Therefore, this pumping station is modelled without its flushing function in summer (because it is not possible to model the demand from Delfland, and because of the relatively small volumes), and with pumping at peak events in summer and winter.

Gouda inlet is for flushing the system in the summer, and is therefore not modelled in predictive mode: the measured data is taken as input.

The shiplocks are regarded as being closed, because the regular small volume that comes in to the basin has already been taken into account in the fixed inflows.

So in short, summer flushing is not modelled, because also in an operational setting this is done according to schedule and will not be done during excessive rainfall events. All control structures that do have a discharge function during excessive rainfall events are modelled to show that the model can accurately reproduce past critical events and is suitable to be used for predictions. In operational mode, the model could be updated every decision time step with the measured flows. All relevant discharge and inlet structures are monitored in real-time with a telemetric system. The Aquarius software can be controlled automatically by external modules, which allows full incooperation in an online decision support system. This would ensure that the model keeps an accurate initial state, during summer and winter.

Calibration of the final model

With these settings the final model is calibrated by adjusting the soil outflow resistances. Note that varying the soil outflow resistances does not have a uni-directional effect, meaning that lower resistances do not always result in steeper peaks, and that higher values do not always result in flatter peaks. This behaviour of the soil outflow resistance is because different soil layers, with different outflow resistances are used. Therefore, increasing soil outflow resistance in one layer may force the groundwater level into the upper layer, where lower soil outflow resistances may be used. So, with calibration, a wide range of soil outflow resistances should be checked with intervals that are not too big (20 days), for a wide range of events, to find the best interval; then the optimum can be found with even smaller increments.

The final calibration results are presented in Figure 4.12 to Figure 4.15. Figure 4.12 shows that the monthly discharge volumes of the year 2002 have been improved considerably with respect to the first model (Figure 4.7). Note also the accurately modelled total yearly volume (0.7% error) for 2002, while only the total volume over 6 year simulation (1997-2002) was calibrated with the sine functions and fixed flow adjustments. The modelled volumes of the 2000 and 2001 events now match the measured volumes very well (Figure 4.13).

Figure 4.14 shows that the improved representation of slow processes and monthly and seasonal volumes, has also resulted in an improved water level modelling of the Rijnland storage basin. The 2000 event is now calibrated more accurately (Figure 4.14) as compared to the first model (Figure 4.5). The results for normal flow periods remain the same (Figure 4.15).

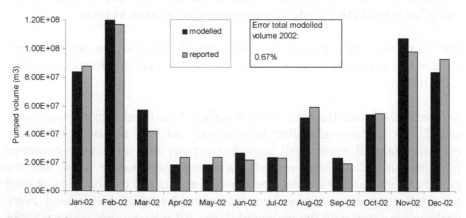

Figure 4.12 Monthly pumped discharge volume from Rijnland storage basin in 2002 of the final model. Note the improvements in summer months and October and November compared to Figure 4.7. Note also the accurately modelled total yearly volume (0.7% error), while only the total volume over 6-year simulation (1997-2002) was calibrated.

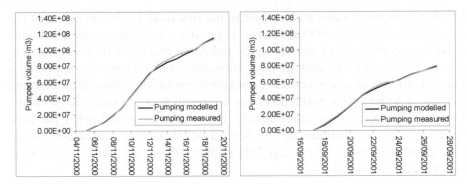

Figure 4.13 Cumulative pump discharge volume for events in the wet winter season in 2000(a) and 2001(b) after external modelling of unknown processes and calibration.

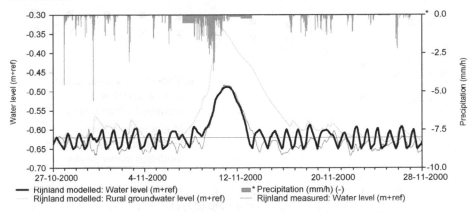

Figure 4.14 Calibration of the event of November 2000, after the unknown processes had been included as external data driven models. The modelling of the peak has improved considerably with respect to the first model (Figure 4.5)

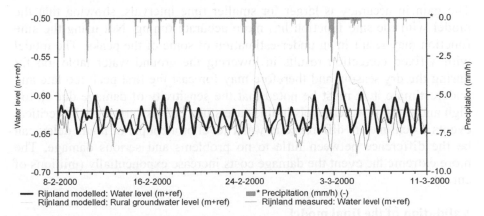

Figure 4.15 Model results for a normal flow period in February 2000, after the unknown processes had been included as external data driven models. There are not many differences with the first model (Figure 4.6)

While the improvements with respect to the first model, after inclusion of the sine function and re-calibration, are large and clear, the question remains what the benefits are of modelling the unknown processes (in this case with a sine function) compared to only correcting the volumes by applying a fixed, constant flow correction. Comparative analysis shows that although the use of a fixed flow correction does not result in consistantly large errors, the overall accuracy is considerably less than the model with the sine function. This is illustrated by the Nash-Sutcliffe (1970) coefficients of the two models, derived for the top 20% of the inflow data, for different time intervals (Figure 4.16).

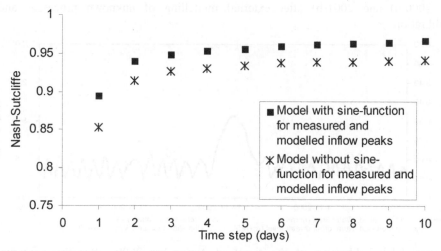

Figure 4.16 Nash-Sutcliffe coefficients for the Rijnland model, with the sine function and with a constant flow correction, for different time intervals.

The gain in accuracy is larger for smaller time intervals, showing that the model with the sine function has more accurate timing. Not using the sine function may result in an under-estimation of some of the peaks. The model with a fixed correction results in lowering the ground water table too far during the dry season, and therefore may forecast the first peak too late and underestimate it. It must be noted that the sensitivity of damage due to too high and too low water levels in this case study area is very high. For critical events 5 centimeters difference in water level or a couple of hours delay can be the difference between little to no problems and serious damage. The more extreme the event the damage costs increase exponentially (millions of euros).

Validation of the final model
The final model has been validated for the years 2003 and 2004. Figure 4.17 shows that the cumulative pumped discharge volume from Rijnland is now modelled very well for the calibration period (compared with the first model

result of Figure 4.8) and remains accurate during the validation in 2003 and 2004. The modelled volumes of two validation events match the measured volumes well (Figure 4.18). Note that small differences here can be the cause of expert based deviation from the operational routine by water managers, which shows up in the measured discharge volume.

Validation results are particularly successful, when considering that 2003 was an extremely dry year in which exceptional control measures have been taken. While part of these measures are included in the input data, it is

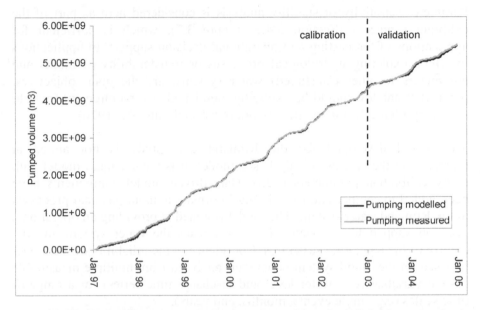

Figure 4.17. Cumulative pump discharge volume from the Rijnland storage basin during calibration and validation. Note that the modelled cumulative volume now matches very well compared to the first model (Figure 4.8) and that the development of cumulative volume remains accurate during the validation period of 2003 and 2004.

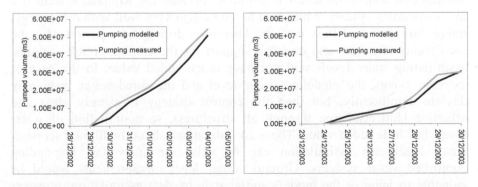

Figure 4.18. Cumulative discharge volume for validation events in January and December 2003.

expected that the validation results would have been even better if all data from exceptional inlets would have been available.

These results show that the model can be used for operational decision support for warnings of peak storage basin water level events, and for providing predictions of discharge volume.

4.4.8 Discussion

Parameter sensitivity or stability analysis is considered here as part of the calibration process (Section 3.4.2, Figure 3.7), which is sufficient for applications of forecasting and operational decision support. In applications in which exploring hydrological processes or transferability of model and parameters to other (ungauged) water systems, are the main objectives, parameter stability should be explicitly mentioned as a requirement (next to accuracy, Figure 3.7) in the framework (Nash and Sutcliffe, 1970).

The model developed for the Rijnland case study is not aimed at representing the best physically based model. It is a conceptual model with many calibration parameters (>25). Data driven model components have been added to compensate for missing information about physical processes in the Rijnland water system. The model is aimed at providing an operational decision support tool, which reliably simulates the water system and the control of the regulating structures. Confidence in the reliability of the model is based on the sound system behaviour, analysed from multiple measurable system variables (e.g. water level and discharge time series) for a range of time scales (e.g. single events, months, and years).

In operational support of water system control there is often the possibility to include data assimilation schemes in the modelling process. In the Rijnland case, in operational use, the only measured data that will be used is precipitation, which results in model updating, not in data assimilation. Data assimilation with water levels is not done, because the Rijnland system is a fast responding system, which can be controlled very well within the target range during normal flow periods. Therefore, during these periods, for the predictions of inflow and decision support for the coming control actions, assimilating water levels will not have much added value. In the case of extreme events, the modelled water level and measured water level may deviate significantly, but then the control strategy is already clear and effective (full discharge through all structures), so assimilation of water levels has little added value. These considerations about the limited scope of water level data assimilation can be valid for many fast responding controlled water systems. Ground- water level measurements would be valuable to improve the model's initial state by data assimilation, however these measurements are currently not operationally available to Rijnland.

The challenges described, and the proposed modelling framework are relevant for many operationally controlled water systems in the Netherlands, because of the historic development of the use of models. Until recently the focus was on water level simulations for design and scenario analysis, in which case the results of the first model could well be satisfactory. Now models are more and more used for prediction and decision support for control structure operation, and water boards have to consider whether the existing model performance is still sufficient and whether it is worth the effort of solving unresolved issues, such as long term volume balances. The presented framework for water system control modelling (Figure 3.7) in many cases can be implemented within a reasonable amount of time (e.g. only one additional iteration). Using this approach, solving or at least modelling the unresolved long-term issues, does not only increase the understanding of the water system, but has also shown how to improve the short term water level and discharge predictions that are so important for operational decision support.

4.5 Ensemble forecasts verification

To asses the reliability of the forecasting product for use in AWM (step 4 in the AWM framework, Figure 3.12), a verification analysis of 7.5 years has been performed with an ensemble precipitation archive and water level hindcasts.

The archive and hindcasts were analysed for the period from 25 April 1997 to 31 August 2004.

4.5.1 Precipitation ensemble forecasts archive

The precipitation forecasts are the ECMWF EPS precipitation forecasts for location De Bilt in the Netherlands. The 6-hour precipitation amounts were accumulated according to the different precipitation thresholds used in the experiments. For example, the first experiment compares the forecasts and the measured precipitation for a 15 mm day-1 threshold. Therefore, for each 6-hour time step, the forecasted precipitation for the past 24 hours was accumulated to give forecasted precipitation per day.

4.5.2 Water level hindcasts

For the water level hindcasts an Aquarius water-system control model was used. The water level hindcasts were determined by feeding the ECMWF EPS precipitation forecasts into the water-system control model. The model was applied deterministically, without accounting for additional sources of uncertainty. Real-time forecasting was simulated using a spin-up period of 30 days with area averaged measured precipitation. Following this spin-up period, the 50 ECMWF EPS precipitation members were fed to the deterministic Aquarius model to generate the ensemble forecast of water levels up to 10 days ahead.

Because the model contains the routine operational strategy, these forecasts show when the current control strategy is not sufficient to prevent high water levels, and therefore, when anticipatory control is needed.

4.5.3 Event based verification for water managers

An analysis method was chosen that best fits the needs of the operational water managers. An analysis of the precipitation and water level hindcasts for a period of 7.5 years was done to enable the water managers to verify the analysis steps and results according to their own experience and measured data. A verification tool was developed in which the user can define the variables of a threshold based decision rule for EPS, being event threshold (precipitation or water level threshold), forecast horizon and probability threshold (minimum forecasted probability that the event threshold will be exceeded).

The main concern of the Water Board is to limit damage due to floods. The water managers want to know how many critical events, which can be handled only by anticipatory control, will be forecasted by the system (hits). Since unnecessary flood-control actions, such as water release and pumping, may cause damage, the Water Board is also interested in the number of false alarms.

The decision rule defines the event threshold (precipitation or water level), the forecast horizon and the probability threshold. For every evaluated combination of forecast horizon and probability threshold, the corresponding forecasted value is determined using the percentile function (probability threshold 0.1 corresponds to the 90th percentile of the ensemble members). The forecasted value is compared with the event threshold. When the forecasted value exceeds the event threshold, the date and time are marked as a forecasted event.

The comparison between measured and forecasted events was done for the time at which a critical event begins. The forecast of the beginning of an event is important to allow for effective anticipatory control actions. The forecasted time at which the event begins should be within a predefined range (e.g. one day) of the actual beginning of the event. For every measured critical event, it was determined whether it was forecasted (hit) or missed

(missed event). For every forecasted event, it was determined whether the event actually occurred (hit) or not (false alarm).

4.5.4 Precipitation and water level thresholds

The numbers of hits, missed events and false alarms over a period of 7.5 years were determined for three different precipitation thresholds and one water level threshold. For each precipitation or water level threshold, forecast horizons up to nine days and the full range of probability thresholds were analyzed.

4.5.5 Presently used precipitation threshold for anticipatory pumping

The Rijnland Water Board currently applies a precautionary threshold-based decision rule, with a precipitation threshold of 15 mm day-1 and a 1-day forecast horizon. In order to test the possibility of extending the forecast horizon from one day to three days or more, the 7.5 years of ECMWF EPS forecasts and measured precipitation were compared for the 15 mm day-1 precipitation threshold. Eighty-five events exceeding this threshold were identified (the sample climatology is 0.03). Of these 85 measured events, 78 could have been anticipated using a forecast horizon of three days and taking the highest of the 50 forecasted precipitation values as probability threshold. This anticipation of events would have been done at the expense of 352 false alarms. For a period of 7.5 years, this means that on average once a week there would be an alarm and approximately five out of six of these alarms would be false. Fewer false alarms can be achieved by applying decision rules with higher probability thresholds, but this also reduces the number of hits.

Figure 4.19 shows this relationship between the decision rule and the number of hits and false alarms, for the fixed precipitation threshold of 15 mm day-1 and a varying forecast horizon and probability threshold. It shows that the number of hits decreases with increasing probability threshold. For probability thresholds greater than 0.04, the number of hits also decreases with increasing forecast horizon. The number of false alarms decreases with increasing probability threshold but increases with increasing forecast horizon for probability thresholds up to 0.07. For probability thresholds greater than 0.10 the number of false alarms decreases with increasing forecast horizon, indicating that for long forecast horizons the system is not forecasting the event with high probability. Figure 4.19 also shows that the decrease in the number of hits and false alarms with increasing probability threshold increases for longer forecast horizons.

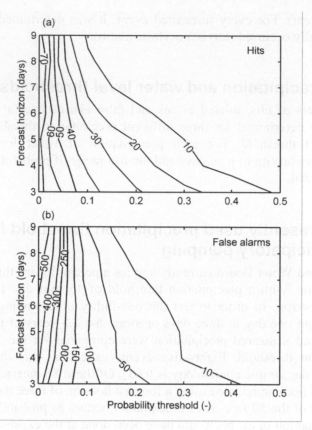

Figure 4.19 Contours of number of hits (a) and number of false alarms (b) of the ECMWF EPS precipitation forecasts for 85 precipitation events in the Rijnland water system of 15 mm day-1 or more. (ECMWF EPS precipitation for location De Bilt, from 25 April 1997 to 31 August 2004)

The benefit of Figure 4.19 is that the Water Board can see the performance of a range of possible decision rules at a single glance. The data behind Figure 4.19 is also directly accessible with the verification tool, enabling further analyses of specific combinations of forecast horizon and probability threshold. Figure 4.20a shows the number of false alarms and missed events, together with the forecasts that were too early, for the 3-day forecast horizon and a range of probability thresholds. It can be seen that for the lowest probability thresholds some events were forecasted too early. This indicates that when looking only at the maximum ensemble members the forecast may overestimate the amount of precipitation. Figure 4.20b shows the actual forecasts for a 3-day forecast horizon and a 0.04 probability threshold (96th percentile). This combination results in good forecasts of dry weather periods. The precipitation events also show good agreement, but most of the forecasts overestimate the precipitation. This can result in false alarms, as is the case for 10 September 2003 (Figure 4.20b).

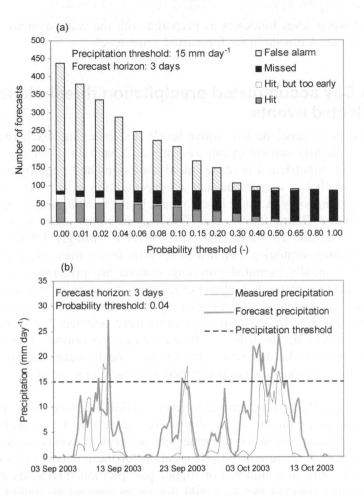

Figure 4.20 Detailed analyses of performance of threshold-based decision rules with ECMWF EPS precipitation forecasts. (a) Number of hits, events that have been forecasted too early, missed events and false alarms for a 15 mm day-1 precipitation threshold and a 3-day forecast horizon. (b) Measured daily precipitation and forecasted daily precipitation for a 3-day forecast horizon and a probability threshold of 0.04 (96th percentile).

The number of hits of 15 mm day-1 events may be good, but the number of false alarms is high. Furthermore, the Water Board knows that in the analysis period only a few critical events occurred, not 85 as indicated, with the precipitation threshold of 15 mm day-1. The Water Board uses this low event threshold as a precautionary measure to account for the uncertainty of the weather forecast. In reality an operational water manager takes into account the measured precipitation of the previous days and the present water levels, because the initial conditions of the water system determine whether an additional 15 mm will cause flooding problems or not. There are two ways to account for initial conditions. One is to determine precipitation thresholds over a number of days (e.g. 30 mm per 3 days). The other is to look at the water level hindcasts as prepared with the water-system control model and to use a threshold for forecasted water level.

4.5.6 3-Day accumulated precipitation threshold for selected events

Measurements of precipitation, water levels and pumping discharge were analyzed to identify critical events (Figure 3.12, step 1). Looking only at precipitation is insufficient because initial conditions and pumping strategy determine whether a certain amount of precipitation results in a critical situation. Looking only at water levels would be the most logical thing to do, since the Water Board knows that flooding problems start occurring at certain water levels. However, the water level is strongly influenced by variable pumping strategies. For instance, water levels may rise during the night when manually operated pumping stations are preferably not used. This could lead to relatively high water levels in the morning that are easily reduced once the pumping stations are put into action. To avoid this type of event being characterized as critical, events were selected that resulted in high water levels for at least 12 hours despite continuous pumping at maximum capacity. In this way, nine critical events were identified. All further verification analysis was performed with these nine events.

To analyze the effect of accumulating precipitation forecasts over time, 3-day accumulated precipitation thresholds were determined. An analysis of the measured precipitation for the nine selected critical events showed that the minimum 3-day accumulated precipitation for these events is 40 mm in winter and 45 mm in summer. For higher precipitation thresholds at least two of the nine selected events would not be recognized according to the measured precipitation.

To compare the results of the precipitation forecasts with the water level forecasts, a water level threshold was determined for the same nine critical events. Water levels that were modelled using the measured precipitation as input, were compared with the selected events. A water level threshold of -0.57 m+Ref during winter and -0.55 m+Ref during summer would help to identify the nine selected events.

Using these precipitation thresholds and water level thresholds, the performance of the full range of threshold-based decision rules for ensemble forecasts was determined as in Section 4.5.5. Of the nine selected events, seven could have been forecasted using the precipitation thresholds, and six could have been forecasted using the water level thresholds. Figure 4.21a and Figure 4.21b show that fewer events are successfully forecasted for the short forecast horizons (three and four days ahead). This could show that the ECMWF EPS forecasting system has been optimized to provide good probabilistic forecasts for three to five days. During the first two days, the disturbances in the initial conditions have not grown enough to present the full spread of possible events. 3-Day accumulated values of the 3- and 4-day forecast horizon include the first two days and may therefore underestimate the amount of precipitation. Figure 4.21c shows that with the precipitation threshold of 40 mm per 3 days in winter and 45 mm per 3 days in summer, the number of false alarms is decreased substantially compared to the threshold of 15 mm day-1 (Figure 4.19b). Figure 4.21d shows that when using the ensemble water level forecasts instead of the precipitation forecasts, even fewer false alarms result. For example, for a 6-day forecast horizon and a probability threshold of 0.1, Figure 4.21c reads close to 20 false alarms, while Figure 4.21d reads less than 5 false alarms.

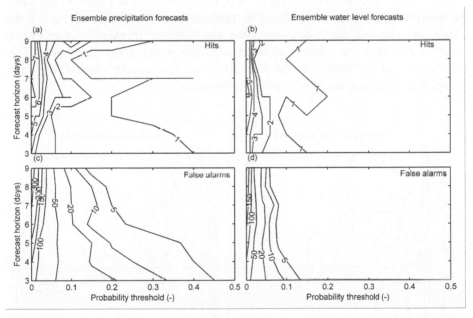

Figure 4.21 Comparison of performance of decision rules based on precipitation forecasts and water level forecasts. [Left] Contours of number of hits (a) and false alarms (c) with ensemble precipitation forecasts for nine selected events. Winter precipitation threshold: 40 mm per 3 days. Summer precipitation threshold: 45 mm per 3 days. [Right] Contours of number of hits (b) and false alarms (d) with ensemble water level forecasts for nine selected events. Winter water level threshold: -0.57 m+Ref for 12 hours. Summer water level threshold: -0.55 m+Ref for 12 hours.

4.5.7 5-Day accumulated precipitation threshold for selected events

To analyze further the effect of accumulating precipitation forecasts over time, a precipitation threshold for 5-day accumulated precipitation was used.

When compared with the measured precipitation, a threshold of 65 mm per 5 days best identifies the nine selected critical events.

When using a precipitation threshold of 65 mm per 5 days, all of the nine events could have been forecasted (8-day forecast horizon, highest forecast value as probability threshold), which is more than the six events that were forecasted using the water level forecasts. The decay of number of hits with increasing probability threshold is stronger for water level forecasts than precipitation forecasts (Figure 4.22a,b). Figure 4.22c shows that the number of false alarms is further reduced compared to the 40 mm per 3 days and 45 mm per 3 days thresholds (Figure 4.21c). The number of false alarms with precipitation forecasts is now comparable to the number of false alarms with water level forecasts. For example, both Figure 4.22c and Figure 4.22d show a maximum of approximately 150 false alarms. Further analyses of the forecasts showed that when using the precipitation threshold, more events were forecasted too early compared with the water level thresholds. This indicates that the timing of the forecasts is more accurate when using water level forecasts instead of precipitation forecasts.

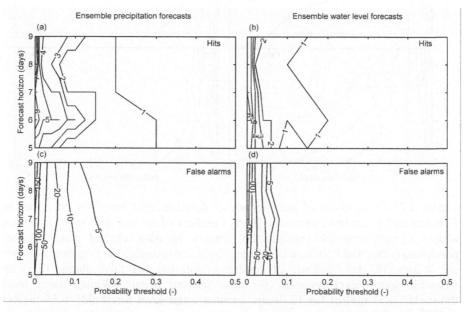

Figure 4.22 Comparison of performance of decision rules based on precipitation forecasts and water level forecasts. [Left] Contours of number of hits (a) and false alarms (c) with ensemble precipitation forecasts for nine selected events. Winter precipitation threshold: 65 mm per 5 days. Summer precipitation threshold: 65 mm per 5 days. [Right] Contours of number of hits (b) and false alarms (d) with ensemble water level forecasts for nine selected events. Winter water level threshold: -0.57 m+Ref for 12 hours. Summer water level threshold: -0.55 m+Ref for 12 hours.

Forecast horizons between 5 and 7 days seem to perform best for forecasting the nine measured critical events. Relative Operating Characteristic (ROC) - diagrams can be used to compare further the performance of these forecast horizons and the precipitation and water level forecasts. For each probability threshold the false alarm rate (probability of false detection, e.g. number of false alarms divided by the number of non-events) and hit rate (number of hits divided by the number of events) are plotted against each other. The points for each probability threshold can be connected to form a curve. The larger the area under the curve (ROC-area) the better the forecast skill (Atger, 2001).

The difficulty in applying the ROC-diagram for a few critical events is that the curves tend to be aligned with the y-axis of the graph, because of the high number of correctly forecasted non-events (an event that is not observed and not forecasted), making the lowest probability threshold decisive for the ROC-area. This problem is reduced in the event-based approach adopted here, by setting the duration of non-events to the duration of false alarms (instead of one day or one forecast time step, as is often applied for the non-event). Figure 4.23a shows that for the precipitation forecasts the 6-day forecast horizon performs slightly better than the 5- and 7-day forecast horizon. The ROC-curves of the water level forecasts for 5, 6, and 7-day forecast horizons are almost the same. Figure 4.23b compares the 6-day precipitation forecasts with the 6-day water level forecasts and confirms that the precipitation forecasts perform slightly better.

When applying day-by-day verification the areas under the ROC-curves (ROC-area) are slightly smaller than in Figure 4.23. The 6-day precipitation forecast ROC-area would be 0.74 instead of 0.86 and the 6-day water level forecast ROC-area would be 0.75 instead of 0.80. These differences are caused by one long precipitation event in 2000 that lasted 4 days and was not forecasted. The contingency table of the 6-day precipitation forecasts (Table 4.1) shows that even the lowest probability thresholds are still higher than the sample climatology of 0.006 (9/1464) and that the decision rule based on one ensemble member exceeding the threshold, leads to many false alarms. Yet, these low probability thresholds with the highest number of hits, are of interest to the Water Board, as its primary concern is to identify critical events.

4.5.8 Discussion

The results show that for the case study of Rijnland Water Board in the Netherlands, the ECMWF EPS precipitation forecasts can be used in flood control to forecast critical events for which anticipatory control actions are needed. The analysis of the different decision rules shows that low probability thresholds (<0.05) should be used to identify critical events.

Forecast horizons between five and seven days seem to perform best for forecasting the nine measured critical events. The most important finding from the present case study is that it is better to use high event thresholds that identify only the truly critical events than to choose low thresholds based on precautionary principles. A thorough event analysis is needed to define these thresholds. The results show that a factor of two to three in the reduction of false alarms can be reached while maintaining the same number of hits (e.g. from 300 false alarms in Figure 4.21c to 150 false alarms in Figure 4.22c).

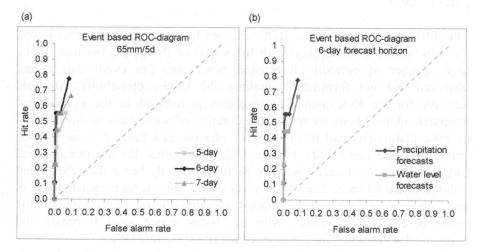

Figure 4.23 Event based ROC-diagrams of ensemble precipitation (a) for 5, 6, and 7 days forecast horizons. For comparison the ROC curves of the 6-day forecast horizon water level forecasts and precipitation forecasts for 65 mm/5 days have been plotted (b). The curves show the relationship between hit rate and false alarm rate for different probability thresholds. The lowest probability threshold is the upper right end of the curves, for the highest probability thresholds the curves reach the origin (no hits and no false alarms).

Table 4.1 Event based contingency table for precipitation forecasts with a 6-day forecast horizon.

65 mm/5days	Observed	Not observed	
Forecast (by at least 1 member)	7	120	127
Not forecast	2	1335	1337
	9	1455	1464

While at first it seems that the use of a water-system control model to translate the precipitation forecasts into water level forecasts results in a reduction of false alarms, this case study demonstrates that a good precipitation threshold can be defined, which results in a similar reduction of false alarms. The precipitation forecasts even forecasted three more critical events than the water level forecasts. However, the timing of the critical events is forecasted better using the water level forecasts.

For practical use, a water-system control model can be used to reduce the risk of false alarms compared to using precipitation forecasts directly. If it is decided to start anticipatory control actions, these can be modelled with the water-system control model. The model shows when sufficient measures have been taken (e.g. extra storage created) and anticipatory control actions can be stopped. When using only the precipitation forecasts, anticipatory control actions will be continued as long as the forecasted precipitation is exceeding the threshold, despite the extra storage that has been created.

The analysis was performed to show how many of the critical events that occurred during the full analysis period could have been identified by the ensemble forecasting systems. Therefore, the effect of seasonal differences and changes due to development of the ECMWF EPS system are not shown. Ideally, for any up-grade of operational forecasting models, a new archive of hindcasts would be created to allow for end-users to adjust their decision rules accordingly.

The presented research applied decision rules where a forecast horizon is fixed and this is the only horizon to look at from one decision point to another. When looking at consecutive ECMWF EPS precipitation forecasts they are not always consistent in time. This means that the forecast for day i may show an extreme event coming up, while the forecast for day $i+1$ shows no event at all. Therefore, dynamic decision rules that look at a range of forecast horizons, e.g. three to eight days, may increase the number of hits. In our particular case study, using this range of forecast horizons enables the water level forecasts to identify eight of the nine selected events. Even the last event is recognized by the forecasting system but the beginning of the event is forecasted just over 24 hours too late. Since this can be dangerous for flood-control actions, it is not considered a good forecast. An event that is forecasted too early does not have to be a problem if the adverse effects of prolonged anticipatory control actions are limited. In this case study, forecasts that are too early do not limit the effectiveness of the control actions to reduce flooding problems. Therefore, such forecasts are considered good forecasts in the long-term verification analysis.

The analysis on the basis of only nine critical events has already resulted in clear directions of what are the most effective decision rules. However, a longer analysis period with more critical events would allow for a more detailed definition of the optimal decision rule and with more confidence. Therefore it is important that Water Boards and Meteorological offices generate multi-year (decades) hindcast archives.

The focus of this forecast verification analysis has been on numbers of hits and false alarms, because this enables the Water Boards to verify their absolute requirements in flood protection. Next to these absolute requirements, a cost-loss analysis (Richardson, 2000) is an important factor in the decision of the Water Board whether or not, and how, to apply AWM. Therefore, after new control strategies are developed (Section 4.6), the cost-benefit analysis of these strategies is performed (Section 4.7).

4.6 Anticipatory water management strategy development

The verification results showed a decline in the hit rate with increasing forecast horizon and probability threshold. The water board is interested in the highest possible safety against flooding that can be achieved using the ensemble forecasts. Therefore, an example AWM strategy is evaluated (Figure 3.12) in which anticipatory control is applied if 1 of the 50 ensemble members exceeds the water level threshold. The water level threshold is put to -0.57 m+Ref, because this is a water level that normally occurs only if the inflow into the system exceeds the total pumping capacity of the system (outflow). To provide enough lead-time to lower the storage basin water level before the precipitation event occurs, a forecast horizon of 3 days is applied. To maximise the probability of identifying the critical events, not only the present day forecast (t = 0) is considered for a 3-day horizon, but also the forecasts that were received in the days before (t = -1, -2, -3, -4 days) are considered. The forecast of yesterday (t = -1 day) is evaluated for its 4-day horizon to match with the 3-day horizon of the present day forecast. In the same way the 5-, 6-, and 7-day horizons of respectively the t= -2, -3, and -4 day forecasts are used. Note that because of the risk avers approach adopted here, no relative weights have been given to the older forecasts to express a decline of forecast skill with increasing forecast horizon. If one of the forecasts exceeds the water level threshold, the water level in the storage basin is drawn down to a level between -0.65 and -0.70 m+Ref to create extra storage at the beginning of the event.

The effect of this control strategy was simulated using the combined ECMWF EPS precipitation forecasts and the AQUARIUS water system control model for a period of heavy precipitation in September 1998. One of the ensemble forecasts is shown in Figure 4.24 (an animation of all forecasts for that period can be found in the supplementary pages of the electronic journal paper Van Andel et al., 2008[b]). In addition to the forecasts, the red line presents the measured water level, and the blue line presents the water level as modelled with the measured precipitation as input. The September 1998 peak exceeded the threshold water level of -0.57 m+Ref for about two days. Several members of the ensemble forecast exceed the -0.57 m+Ref level for several days as well. Given the proposed strategy for anticipatory control, therefore, the model will activate the pumping stations and draw down the water level. The results are presented in Figure 4.25.

The red line presents the historically measured water level in the storage basin. The blue line presents the simulated water level with anticipatory lowering of water levels to -0.70 m+Ref on the basis of the EPS water level forecasts. It can be seen that the extra storage prevents the water level from exceeding -0.57 m+Ref. In the periods before and after the event, the modelled and measured water levels are about the same. This shows that for this period no false alarms occurred and no unnecessary lowering of the water level in the storage basin was performed.

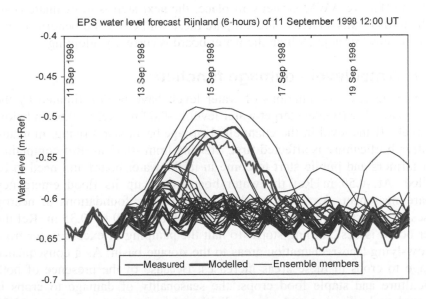

Figure 4.24 Ensemble water level prediction of 11 September 1998 for the Rijnland water system on the basis of ECMWF EPS precipitation forecasts (Van Andel et al., 2008[b]).

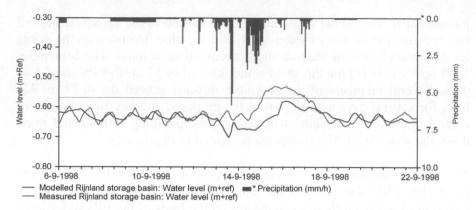

Figure 4.25 Effect of modelled Anticipatory Water Management on water level control in the Rijnland water system. The modelled peak is lower than the measured peak, because in the model the water level is lowered before the precipitation event occurs. Exceedance of the -0.57 m+Ref level is prevented (horizontal line).

4.7 Cost-benefit of selected AWM strategies

With an effective AWM strategy in place, the next step is to evaluate costs and benefits of this strategy and compare these with costs and benefits of the normal control strategy, which the water board is currently adopting.

4.7.1 Water level - damage function

The damage costs as functions of water levels have been estimated by the water board. The (winter) target water level is -0.62 m+Ref ($\approx$ 0.62 m below sea level). If the level in the channelled storage basin starts rising, moisture sensitive horticulture is affected first. Starting from -0.55 m+Ref complaints from farmers and public start to come in to the water board and need to be handled. At -0.50 m+Ref the water board starts up its flood emergency preparedness plan. As water levels rise further the foundations of nearby houses are affected and shipping is hindered. From -0.40 to -0.35 m+Ref the water board issues "milling stops" to halt the pumping of excess water from the low-lying land-reclamation areas to the storage basin. As a consequence damage to crops in these areas increases. Because of the presence of both horticulture and staple food crops, the seasonality of damage to crops is expected to be limited and not explicitly taken into account. From -0.30 m+Ref onward, damage increases rapidly because of damage to crops, houses and infrastructure. With levels of -0.10 m+Ref and higher the stability of the reservoir embankments is affected and flood damage then depends on the location of the first breaches.

The damage costs for too high water levels have been quantified based on maximum cost estimates (in case of inundation) per hectare per land use type (HKV, 2006). The maximum flood damage for Rijnland was then calculated by multiplying the cost estimates with the land use areas in the catchment. The costs related to the different storage basis levels, and duration of their occurrence, have been estimated as percentages of this maximum flood damage by the water board.

When water levels become too low and drop below -0.65 m+Ref, embankments start to be affected, shipping is hindered, and houseboats are damaged. Also complaints from the public come in. Below -0.75 m+Ref the foundations of houses start to be affected and the damage cost increases with further lowering of water levels and their longer duration (Figure 4.26).

The costs of damaged embankments have been quantified based on costs per meter embankment for repairs (1000 euro/m). The length of affected embankments and foundations for different (low) storage basin levels has been estimated by the water board.

For both too high and too low storage basin levels costs of hindrance of shipping have been estimated as a multiplication of the costs per ship per day that is unable to navigate (500 euro/ship/day). Costs of handling of complaints have been estimated as multiplication of labour costs of 500 euro/day).

These estimates are all present worth estimates.

Operational costs can be related to fuel costs for operating pumps (this can be related to the number of pump operating hours), timing of the operation, e.g. pumping at night for manual operated stations requires operators to work at night at extra high tariffs and social costs, also pumping at night causes noise disturbance to nearby households and nature, and number of switching on and -off, which is related to the maintenance of the structures and shortens the life cycle. The operation at night becomes less of a problem since in the Rijnland case study all main pumping stations are going to be operated automatically, also with the pumps increasingly becoming electrically powered and as the housing of pumps is improving the noise disturbance becomes less. There remains only the effect on house boats which, if situated nearby the pumping station, will go up and down with the water level, and costs of more frequent switching on and -off. Since, according to the water board representatives, the present yearly operational and maintenance costs are so low compared to other costs that they are not part of yearly budget planning or strategic consideration, operational costs are left out of the initial cost estimates and of the optimisation analysis.

Intangible costs like human casualties are not taken into account, because these are not likely to occur. Also indirect costs like the reduction of confidence in the warning system are not considered.

The above described damage cost estimates lead to the water level - damage - duration function shown in Figure 4.26. Note that non-linear increase of the costs with prolonged duration of damaging water levels has been taken into account.

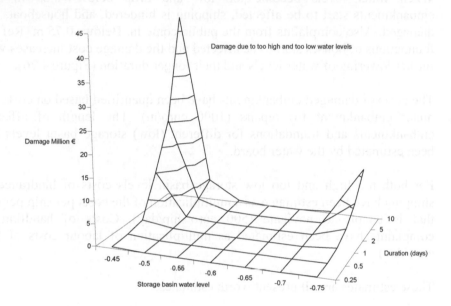

Figure 4.26 Water level-cost function Rijnland storage basin (Van Andel et al., 2009[b]).

4.7.2 Inter-comparison of costs for selected strategies

The Aquarius water system control model of the Rijnland water system is used together with the damage cost function to estimate the total damage cost and flood damage cost of a selection of control strategies for a given evaluation period (7.5 years).

Reference scenario

The reference scenario is the current control strategy. There are two options: to use the measured water levels to estimate the flood and total costs over the analysis period, or to use the modelled water levels with measured precipitation as input. The approach is first to get a reliable and accurate water system control model (Van Andel, 2009[a]), and based on the calibration and validation results to agree on the suitability of the model. Then in the subsequent cost-benefit analysis it is better to use the model results, because human inconsistencies in applying the control strategy and measurement errors in the water levels are filtered out. The simulated water levels are fed to the water level-cost function to calculated the cumulative costs for the simulation period.

Flood risk averse strategy

The flood risk averse strategy developed in Section 4.6 is evaluated as an example Anticipatory Water Management strategy. The flood costs and total costs over the analysis period are compared for the flood risk averse strategy and the reference strategy (Figure 4.27 and Figure 4.28). The results show that the flood damage is reduced considerably by applying the rule based, risk averse AWM strategy, but that the total damage cost (of both too high and too low water levels) becomes higher in this case. Note that presenting the cumulative costs also highlights the important passed events. For example, the strong increase in costs in the year 2000 Figure 4.27 points to the November event that is well known with the water board as a critical event.

Because already with this simple strategy flood damage reduction is considerable, and because the increase of total costs may be seen by the Water Board as investment in reducing flood risk, it was decided to go on to the next phase in the AWM framework; namely, the optimisation of the AWM strategy.

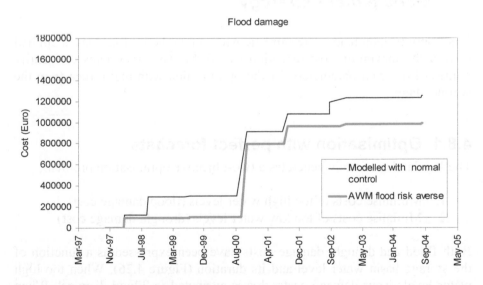

Figure 4.27 Comparison of the flood damage cost estimate of the normal control strategy with a flood risk averse AWM strategy

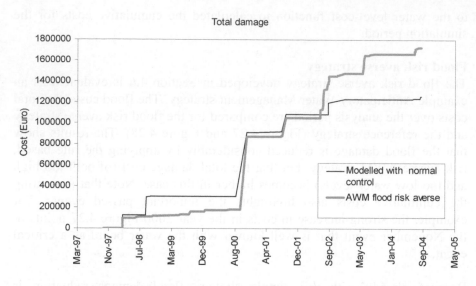

Figure 4.28 Comparison of the total damage estimate of the normal control strategy with a flood risk averse AWM strategy

4.8 Optimisation of Anticipatory Water Management strategy

First, optimisation using deterministic perfect (synthetic) forecasts is applied to find the maximum cost reduction by AWM. This also helps to identify ranges of decision parameters for the optimisation with real forecasts in the second stage.

4.8.1 Optimisation with perfect forecasts

The Rijnland case study generates a two-objective optimisation problem:

1. Minimise costs of too high water levels (flood damage cost)
2. Minimise costs of too low water levels (drought damage cost)

Both flood and drought damage costs have been expressed as a function of the storage basin water level and its duration (Figure 4.26). When too high water levels incur damage costs, this is attributed to 'Flood damage'. When too low water levels incur damage costs, this is attributed to 'Drought damage'.

Perfect water forecasts are prepared by taking the measured precipitation as input to the simulation model of the Rijnland catchment. The lowering of the storage basin level, in anticipation of the simulated inflows, is optimised

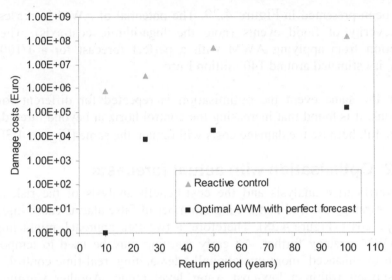

Figure 4.29 Theoretical potential of total cost reduction by applying AWM with perfect (synthetic) forecasts to the Rijnland water system for extreme events with return periods between 10 and 100 years. Note the logarithmic scale of the cost-axis.

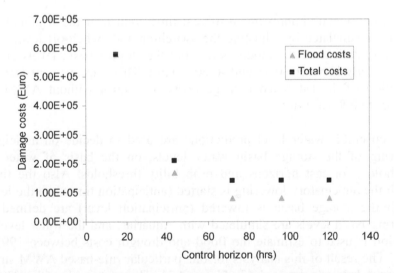

Figure 4.30 Optimisation of control horizon by minimising estimated total damage costs. Both the flood costs and the total costs become stable after 70 hrs. Expansion of the control horizon beyond 70 hrs has no use. The analysis has been performed on the basis of perfect (synthetic) forecasts for a 1/100 year event.

using the genetic algorithm NSGAII (Deb et al., 2002; Barreto et al., 2006). The start-time and end-time of anticipation by pumping with full capacity are optimised for extreme events with estimated return periods between 10 and 100 years (Hoes, 2007). The total costs of these extreme events, both in case AWM is applied and when AWM is not applied (re-active control), have been presented in Figure 4.29. The potential of AWM increases with the severity of flood events (note the logarithmic cost-axis). The cost reduction from applying AWM with a perfect forecast for a 1/100 year event, is estimated around 140 million Euro.

If for the same event the optimisation is repeated for different forecast horizons, it is found that increasing the control horizon beyond three days is not useful, because the damage costs will remain the same (Figure 4.30).

4.8.2 Optimisation with actual forecasts

The verification analysis and the cost-benefit analysis of the risk averse AWM strategy showed that the high number of false alarms cause high total damage costs (Figure 4.28). Therefore, a two-step approach might improve the AWM strategy further. An early warning may be used to temporarily switch to enhanced, more operationally demanding, real-time control, while still staying within a low-cost water level range. Another warning rule should be defined to decide on an AWM action, which temporarily lowers the water level further to a level that induces damage costs, but is considered necessary at the time to limit flood risk.

First a continuous improvement of real-time control of the Rijnland water system is simulated by adjusting the switch-on and switch-off levels of the pumping stations of the model, such that the storage basin levels are kept more strictly between -0.65 and -0.60 m+Ref. This improvement leads to a reduction of the estimated damage costs of control without AWM from $1.2*10^6$ to $0.8*10^6$ Euro.

Then ensemble water level predictions are used to decide on anticipatory lowering of the storage basin water levels, on the basis of water level thresholds, forecast horizons and probability thresholds. Also the time at which the anticipatory lowering is started (anticipation time) and the level to which the storage basin is lowered (anticipation level) are defined. The resulting basin levels are simulated with Aquarius, and the water level-cost function is used to estimate the flood and drought costs between 1997 and 2004. The result of this analysis for one particular rule-based AWM strategy corresponds to one data point in Figure 4.31. The genetic algorithm NSGAII is used to optimise the AWM parameters. Probability threshold, forecast horizons, warning levels, draw-down levels (anticipation levels) and anticipation time are optimised for the wet seasons in the years 1997 to 2004, resulting in strategies with minimum flood or drought damage costs (damage costs of too high and too low water levels). In the summer no anticipation is applied, because the improved RTC would already prevent summer flood damage). The flood and drought costs of all the sampled AWM strategies are plotted in Figure 4.31. The strategies with the least flood damage costs (closest to the y-axis) and with the least drought damage costs (closest to the *x*-axis) make up the Pareto front of optimal AWM strategies.

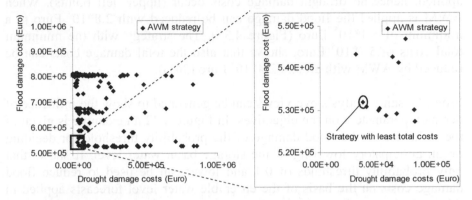

Figure 4.31 Estimatied drought (too low water levels) and flood (too high water levels) damage costs for AWM strategies generated with NSGAII optimisation. Costs are evaluated for the period between 1-9-1997 and 24-4-2004. The lower-left corner of the Pareto front shows strategies with total cost reductions of around 2.4*105 Euro compared to strategies without anticipation (Van Andel et al., 2009[b]).

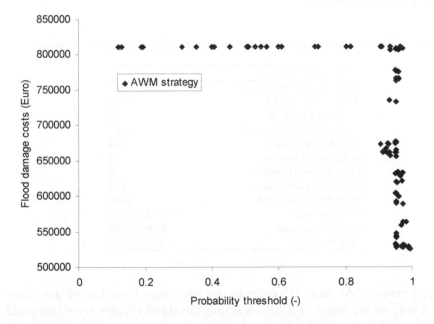

Figure 4.32 Estimated Flood damage costs versus Probability threshold for the 150 least total cost AWM strategies determined by NSGAII optimisation. Costs are evaluated for the period between 1-9-1997 and 24-4-2004. The graph clearly shows how flood damage reduces when low probability thresholds are applied. Meaning that when only 1 or 2 ensemble forecast members are required to exceed the warning level, then most critical events will be identified and the forthcoming damage reduced by AWM strategy.

It can be seen in Figure 4.31 that the maximum estimated flood damage is $8.1*10^5$ Euro. This corresponds to the situation where no anticipation is applied, hence no drought damage costs occur (upper, left points). When AWM is applied the flood damage can be reduced with $2.8*10^5$ Euro to a minimum of $5.3*10^5$ Euro (Figure 4.31). The strategy with the minimum total costs of $5.7*10^5$ Euro, shows that also the total damage costs can be reduced by AWM with around $2.4*10^5$ Euro (30%).

From the same analysis overviews can be generated to show the influence of decision parameters on the objectives. In Figure 4.32 an example is given of the influence on the flood damage of the probability threshold for deciding on an anticipatory lowering of the storage basin water level. It shows that only probability thresholds of 0.1 and lower can be used to reduce flood damage costs on the basis of the ensemble water level forecasts applied in this analysis.

Figure 4.31 and Figure 4.32 show how optimisation can help in selecting decision variables in the rule based AWM strategies. For this case study, for this analysis period, the least-cost strategy found consists of the following decision parameter values:

Table 4.2 Optimal decision parameter values and minimum damage estimation

Decision parameters	
Water level threshold A	-0.57
Anticipation level A	-0.66
anticipation time A	25
Water level threshold B	-0.47
Anticipation level B	-0.74
Anticipation time B	5
First forecast horizon	76
Last forecast horizon	139
Probability threshold	0.05
flood damage	€532,000.00
drought damage	€38,000.00
total damage	€570,000.00

The denotations "A" and "B" refer to the two-stage lowering of the water level. First, on the basis of a low warning threshold (Water level threshold A; -0.57 m+Ref) the level is lowered, pre-cautiously, to Anticipation level A of -0.66 m+Ref. This level causes only limited damage costs. If water level forecasts also exceed the warning threshold, Water level threshold B, then the water level is lowered further to Anticipation level B, -0.74 m+Ref. The forecast horizons included, range from 3 to 6 days. This is consistant with the results of the verification analysis. The anticipation time A, however, is only 1 day, while optimisation with perfect forecasts indicated that 3-days would be needed for extreme events. The reason that this did not come back in the optimisation with real forecasts is that none of the real forecasts managed to predict the extreme event of November 2000. Hence, the AWM strategy is optimised to the smaller events that could be well predicted. For flood risk averse strategies it would therefore be needed to apply apply a longer anticipation time. This shows that for the adoption of the AWM strategy the results of the optimisation analysis should be handled with care.

4.9 Adoption of AWM in operational management policy

AWM can be applied in Rijnland both for the reduction in the costs of too high water levels, and of the total damage costs. The analysis period needs to be expanded in order to generate more reliable optimal control strategies. The rule-based AWM strategy is satisfactory for this case study.

Because the cost-water level function is for a large part estimated from country average unit flood damage estimates and expert judgement from the water board, the absolute cost estimates are to be considered as indicative. Verification of the estimated costs is difficult, because most of the components are hidden costs in the sense that they are usually not actually determined and declared. Only for the extreme range of the water levels, with dike breach or inundation of the polders, the direct damage costs are analysed, e.g. for other Water Boards in the Netherlands in 1998 and 2000. When monitored also the cost estimates for the less extreme water levels could be validated and further detailed and improved.

As alternative to AWM, further improving the real-time control can be investigated as well. The timing of the switching-on of pumping stations at the beginning of the event is crucial. The water board could already reduce the risk of flood damage by starting to pump earlier with all pumping stations at full capacity, instead of a stepwise approach of starting up the pumping stations. Because this strategy would lead to a more frequent switching on and off of the pumps with short intervals, it may not be a preferred strategy in terms of operational costs (electricity, maintenance costs). It can, however, be used as a strategy in combination with flood risk averse warnings. Even if a large part of the cost reduction can also be achieved by further improving the RTC, AWM is still preferable because it reduces the risk of flooding.

An additional incentive to somehow include medium-range ensemble precipitation forecasts in the operational water management, is that water authorities will increasingly be held accountable for having used all the information available. In todays information society individual citizens, when confronted with water damage, will look up the weather forecasts of the days before on the internet and ask the water authorities why they did not anticipate the forecasted rain event.

If the water board decides to implement AWM, it is expected that the rule based AWM will easily fit within the current legislative structure of operational water management, because the Principal Water-board of Rijnland has already started applying AWM on the basis of rainfall forecast thresholds. A decision support system, including deterministic weather forecasts, a rainfall runoff-model, and a decision model for control of the pumping stations is already in place and operational. This would only have to be expanded to include the real-time streaming of the ensemble forecasts and the AWM strategy proposed. Then, while running the forecasts in parallel (off-line), operational water managers can familiarise themselves with AWM, the verification and cost-benefit analysis periods can be continued, and the operational reliability can be assessed.

5 Case study 2 - Upper Blue Nile

5.1 Introduction

While the countries of the Nile may be mostly known for their droughts and subsequent famines, the basin also faces frequent flooding problems. Mainly the Blue Nile river in Ethiopia and Sudan, up to Khartoum, overflows its banks (almost) every year. Despite this recurrent problem, few flood forecasting and early warning systems are in place. This has partly been attributed to the limited data exchange among the riparian countries before the Nile Basin Initiative (NBI) was established. Now it is possible for researchers within the relevant disciplines to function effectively in Nile flood management projects. This case study focuses on flood forecasting and early warning for two sub-catchments, Ribb and Gumara, of Lake Tana in the Upper-Blue Nile in Ethiopia (Amare, 2008). These areas have been identified as two of the target areas for the Flood Preparedness and Early Warning project of the Eastern Nile Technical Regional Office (ENTRO).

Lake Tana can be considered as the source of the Blue Nile. The lake is surrounded by sub-basins with a total area of 12000 km^2. The surrounding sub-basins are drained by several small streams and 11 major rivers that flow into the lake. The eastern portion of the basin is drained by the Ribb and Gumara rivers that account for 28% of the basin area (Kebede, 2006). These two rivers flow to the lake passing through the flat fields of the Fogera flood plain (Figure 5.1).

5.2 Problem description

Flooding is not new to Ethiopia. Floods have been occurring at different places and times, with varying, but often with manageable or 'tolerable' severity. In recent years, however, the country has been threatened by more extreme flooding and severe damage. Most of these flood disasters are attributed to rivers that overflow or burst their banks and inundate downstream flood plains, following torrential rains in the upstream highlands, with duration of several days.

The flooding problems of Ribb and Gumara rivers are of similar nature. The river flow increases from continuous rainfall on the upstream part of the catchments and local rainfall on the flood plain. Areas in the Fogera flood plain that are most at risk from flooding are located between these two rivers. During high floods, people have to live in chest-high water levels, roads become impassable and communication between affected people gets limited to swimming. Fogera is in an administrative district (a Woreda). The Fogera Woreda comprises a land area of 1095 km^2, and has a total population of 243000 people (SMEC, 2006). Within Fogera, 6 or 7 sub-districts are particularly flood-prone. This amounts to approximately a quarter of the Woreda land area.

In spite of the recurrent flood problem, the existing disaster management mechanism is primarily aimed at strengthening rescue and relief arrangements during and after major flood disasters. No decision support systems and anticipatory management strategies to mitigate the flood damage are present. Regional and national flood management authorities (ENTRO) want to research the potential of flood forecasting and early warning for mitigation measures.

Because of limited financial resources, as an additional requirement for the forecasting system, it was to be composed of free and open source weather forecasting and hydrological modelling products.

5.3 Data

5.3.1 Geographical data

The digital elevation model (DEM) data of the Shuttle Radar Topographic Mission (SRTM) was used (SRTM, 2008). The DEM has a spatial resolution of 90 by 90m at the equator.

A soil data set, following the FAO classification, and land use and land cover datasets were obtained from the Ministry of Water Resources of Ethiopia.

5.3.2 Meteorological data

The meteorological data was provided by the National Meteorological Agency (NMA) of Ethiopia. Data from some of the stations, which were not available at the NMA, were collected from the Bahirdar metrological office. The locations of the meteorological stations is shown in Figure 5.1. A period of seven year (2000-2006) was used for analysis.

The rainfall data is daily total rainfall. Most of the stations exhibit significant gaps. Hamusit station is excluded from analysis because 24% of the data is missing. The gaps for most of the stations lie in the rainy season of the year, when it affects the results of hydrological simulations most. For example one of the rainfall time series is shown below (Figure 5.2).

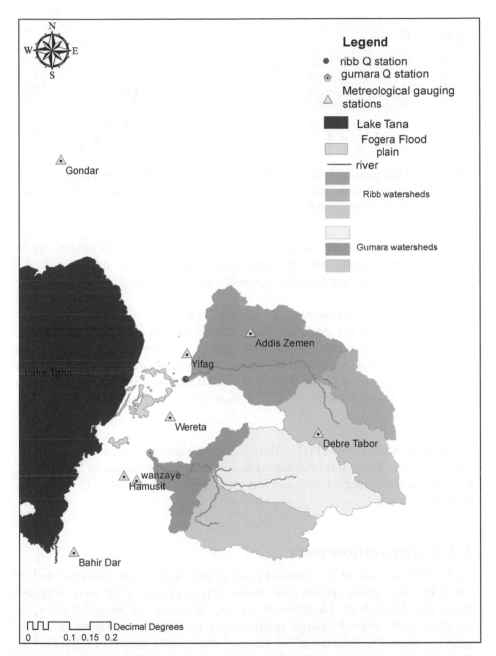

Figure 5.1 Ribb and Gumara catchments with the locations of hydro-metrological gauging stations.

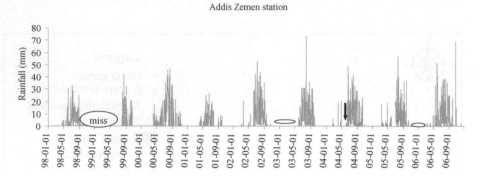

Figure 5.2 Data gaps of Addis Zemen gauging station, indicated by circles for long periods of missing data and an arrow for a short period of missing data. In the analysis only the data between 2000 and 2006 was used.

In addition, no rainfall station exists in the Gumara catchment, while neighbouring stations exhibit significant data gaps in similar periods. Therefore, satellite based rainfall estimates are used to fill the data gaps.

Tropical Rainfall Measurement Mission (TRMM) datasets are freely available through the National Aeronautics and Space Agency (NASA, 2008). The datasets provide the opportunity to have rainfall estimates in regions where conventional rainfall data are scarce (Kummerow, 2000). This study makes use of the daily $0.25°$ x $0.25°$ TRMM and other rainfall data set (3B42 V6) from 2000 to 2006 for four pixels covering the study area (NASA, 2008).

Piche evaporation data (PET) from Bahirdar station is used. The monthly data from this gauge has been correlated with long time series of PET data at Gondar airport station to convert the PET evaporation to Potential evaporation.

5.3.3 Streamflow data

Daily flow records of the Gumara river at the station near Bahirdar, and of the Ribb river at the station near Addis Zemen (Figure 5.1), were obtained from the Hydrology Department of the Ministry of Water Resources, together with stage-discharge relationships and river cross-sections at the gauging sites. The location of the gauging stations are shown in Figure 5.1. Data was available from 1998 to 2006, which covers the analysis period (2000-2006).

The data of the Ribb River was almost complete with only seven days missing, except for the year 1998, where peak discharges are missing (Figure 5.3). The data for the Gumara river, however, has gaps, which are all observed to occur during the dry period of the year. Therefore, the recession curve method is used to fill the gaps (Maidment, 1994). The complete time series of both rivers are plotted in Figure 5.3. It can be observed from this plot that the Gumara River, while having a smaller catchment area, has more discharge in all seasons than the Ribb river.

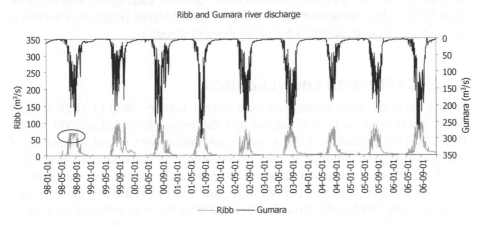

Figure 5.3 Gumara and Ribb daily river discharge from 1998 to 2006. The discharge data of the Ribb river in the wet season of 1998 seems too flat (circled), pointing to measurement errors. The data between 2000 and 2006 was used for analysis.

5.4 Hydrological model

The Hydrologic Engineering Center-Hydrologic Modelling System (HEC-HMS) is a physically-based semi-distributed model (USACE, 2003). It is designed to simulate the rainfall-runoff processes of dendrite watershed systems. The software is freely available. It has been selected for use in this study partly because it has been tested in Upper Blue Nile, and has resulted in good performance (Bashar and Zaki, 2006).

5.4.1 Model set-up

The SRTM 90m DEM was used for catchment delineation. The HEC-HMS Soil Moisture Accounting model was used to allow for continuous simulation. For the direct run-off computation the Clark unit hydrograph method was used. The linear reservoir was adopted for base flow calculation methods, because this module is suitable with the soil moisture accounting model (USACE, 2003). The Muskingum method is used for flood routing in this study for the reason of data limitation to employ the conceptual kinematic wave model.

The Ribb and Gumara catchments have been modelled separately, each with three sub-catchments (Figure 5.1). The area-average rainfall was estimated by making use of the gauge weighting method. The first estimate of the gauge weights was made by making use of the Thiessen polygon method. The Thiessen polygons ware not used alone because of the scarce distribution of the gauges, especially for Gumara catchment, where more than 50% of the catchment lies outside of the Thiessen polygon. Therefore, gauge weights were adjusted based on expert judgement.

5.4.2 Calibration and validation

In this study the daily streamflow data from 1 January 2000 to 31 December 2003 has been used for calibration and the period from 1-Jan-2003 to 31-Dec-2005 for validation. First, manual calibration with visual inspection of the measured and monitored streamflow data was performed to provide a good estimate of the parameters. Then, automatic calibration was applied using the Peak-Weighted Root Mean Square Error (PWRMSE) and Volume Percent Error (VPE) objective functions. PWRMSE was selected as it gives greater overall weight to errors near the peak discharge without significantly affecting the VPE calibrated parameters. Both the default hard-constraints, which limit the range of parameter values within reasonable physical intervals, and soft-constrains on the basis of physical implications of the parameters, were used to limit the range of possible values. The Univariate Gradient search method was used.

Figure 5.4 and Figure 5.5 show a reasonable fit for the calibration of both the Ribb and the Gumara model. The validation of the Gumara river (Figure 5.6 and Figure 5.8b) shows that the model over-predicts, but the trend is reasonable and over-prediction for flood warning applications can be considered positive from a flood risk averse approach. The validation of the Ribb river (Figure 5.7 and Figure 5.8a), however, shows strong under-estimation of streamflow for long periods of the wet seasons. The highest peaks of the wet season are captured well.

Seasonality (wet and dry seasons), next to, for example, spatial distribution of rainfall, and soil and land cover heterogeneity, may be an important sources of error in the hydrological modelling. Developing a seasonal parameterisation approach where each simulated year is divided into two simulation periods (wet and dry seasons) and accordingly one parameter set is obtained for each period could be a good step for improvement of the model.

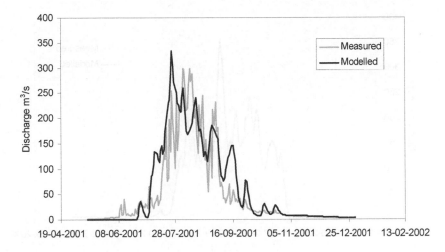

Figure 5.4 Gumara calibration result for 2001

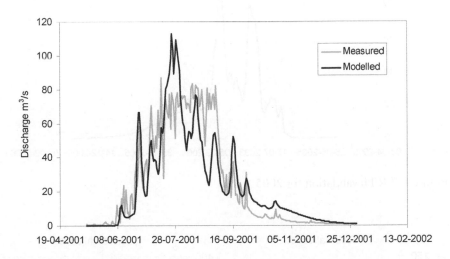

Figure 5.5 Ribb calibration for 2001

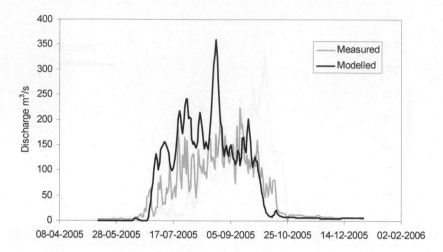

Figure 5.6 Gumara validation for 2005

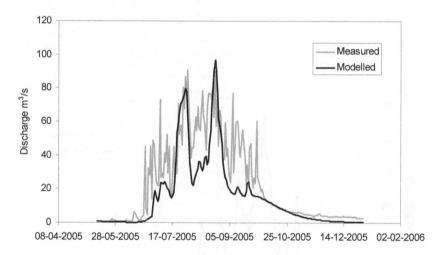

Figure 5.7 Ribb validation for 2005

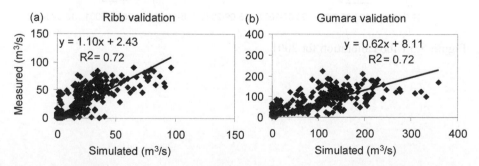

Figure 5.8 Validation results: Ribb correlation (a) , Gumara correlation (b)

5.5 Ensemble forecasts verification

5.5.1 Event selection

Threshold based decision rules can be used for issuing a flood warning. A warning is issued whenever the forecasted flow or water level exceeds a threshold. Thresholds for streamflow need to be related to actual flood events. Flood thresholds for Gumara and Ribb rivers were computed by making use of three different criteria (see also 3.2.1):

-   Recorded flood damages (data from the study area)
-   Flood damages and dates of occurrence from Dartmouth Flood Observatory
-   Bank-full discharge

Each of the three criteria are discussed in the following sections.

Recorded flood damages

The Ribb and Gumara critically high discharges were retrieved by referring to flood damages recorded during the analysis period. Table 5.1 indicates different levels of flood damage in the area. The peak floods in those years are taken as first estimates of the different warning threshold levels.

Table 5.1 Recorded flood damages (DPCC, 2007)

Fiscal year	Affected land area (ha)	Production (kg)	Cost estimate (Eth Birr)	No. of affected Kebeles	Peak floods of the recorded flood years (m^3/s)	
					Gumara	Ribb
2000/01	1566	14,562	2,184,300	5	278	102
2001/02	3697	21,617	2,594,040	7	297	87
2003/04	1155	22,937	3,440,550	3	269	84
2005/06	39	590	118,000	2	223	91

Satellite information

Dartmouth Flood Observatory is an international clearinghouse for GIS data concerning flood inundation, mapped using satellite data. The observatory uses remote sensing to detect, measure, and map major river floods. Information on flood incidences, the dates of occurrence and the damages from the flood can be retrieved free of charge (DFO, 2008). The information for the case study area is presented in map (Figure 5.9) and tabular format (Table 5.2).

It can be observed from the flood map (Figure 5.9) that, on the mentioned dates, the flood extent in Fogera flood plain was the greatest of all the other flood prone areas around Lake Tana. This is further checked with gauge flow records in the corresponding dates and it is found that those dates are recorded with highest flood peaks in that year. A peak of 274 m³/s is recorded on 15 August 2006 at Gumara river. The Ribb hydrograph also indicates continuous peak flows from 90 m³/s to 99 m³/s within the Dartmouth ranges of dates (13 to 18 August 2006).

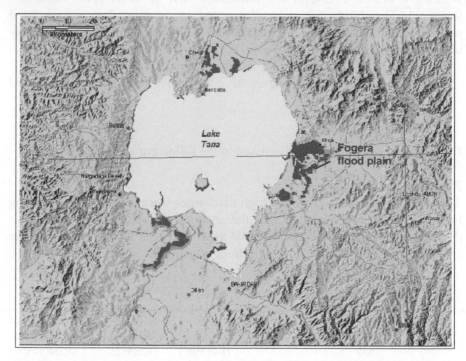

Figure 5.9 Flood areas around Lake Tana sub basin (Aug 13-27, 2006) (DFO, 2008)

Table 5.2 Flood damage in lake Tana sub basin (DFO, 2008)

Areas flooded	Cause of flood	Date of flood	Flood damage
Ethiopia in Blue Nile, Bereka, Ribb, Gumara	Heavy rain	13-Aug to 27-Aug 2006	Ethiopia - 10,000 displaced around Lake Tana, the source of the Blue Nile River. 38,000 displaced by flooding in Amhara region.

River bank-full discharges

One conservative measure of a "flooding flow" is the bank-full discharge. This definition of "flooding" is physically based, but is considered conservative as more than bank full flow is generally needed to cause damage (Carpenter et al., 1999). The bank full discharges and cross-sections at the gauging sites were taken as a reference for the lower warning threshold (Figure 5.10). Based on three criteria discussed, a low, medium and high streamflow warning threshold were determined (Table 5.3).

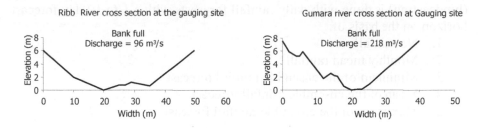

Figure 5.10 River cross-sections of Ribb and Gumara

Table 5.3 Suggested warning threshold

Thresholds (m³/s) /Rivers	Low (threshold 1) (m³/s)	Medium (threshold 2) (m³/s)	High (threshold 3) (m³/s)
Gumara	210	250	300
Ribb	60	85	110

Threshold based flow comparison between Ribb and Gumara rivers
When comparing the peak flows of the Ribb and Gumara rivers it seems that the hydrographs generally follow the same pattern (Figure 5.11). Therefore, getting peak flow warnings for one of the rivers would contribute to flood warnings for the Fogera flood plain. Because of the better performance of the Gumara model and the greater discharges of the Gumara river, it is assumed that the best results for flood warning will be based on the Gumara predictions. Therefore, in the remainder of this case study focus is on the Gumara river only.

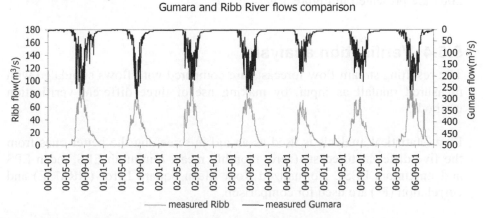

Figure 5.11 Gumara and Ribb river flows comparison.

5.5.2 Ensemble precipitation hindcasts

Five different rainfall forecast archives have been prepared for input to the HEC-HMS model for the period of the years 2000-2006. The re-forecasting (hindcasting) is done with daily rainfall forecasts with a 1 to 10 day forecast horizon on the basis of:

1. No rainfall
2. Monthly mean rainfall
3. Minimum of the ensemble rainfall forecast
4. Mean of the ensemble rainfall forecast
5. Maximum of the ensemble rainfall forecast

As a reference forecast "no-precipitation" is taken. A first improved forecast is prepared by taking the monthly average daily precipitation as input. Monthly mean values are computed from the rainy seasons of the 7 years analysis period (2000-2006).The third, fourth and fifth forecast methods are the Min, Mean, and Max from the ensemble precipitation hindcast archive from the National Centers for Environmental Prediction (NCEP) Global Forecasting System (NCEP, 2008). The ensemble forecasts are freely available on the internet for the entire globe in a grid size of $2.5°$ x $2.5°$. Note that this 'frozen version' of the GFS for hindcasting contains a different model with lower spatial resolution than the currently operational GFS. The 12-hourly ensemble forecast from NCEP consists of fifteen members to a forecast horizon of fifteen days.

5.5.3 Ensemble streamflow hindcasts

The five different rainfall forecast archives are used as input to the HEC-HMS model to produce streamflow hindcasts. For automatically performing the re-forecasting the Hydrologic Engineering Centre Data Storage System Utility Program (DSSUTL) is used. Resulting hindcasts for the wet season of 2001 are presented in Figure 5.12.

5.5.4 Verification analysis

The resulting stream flow forecasts are compared with flows simulated with measured rainfall as input, by making use of three different verification methods.

First, statistical analysis is used as an aid in screening the better ones from the five different forecasts (zero, monthly mean, minimum EPS, mean EPS and maximum EPS). Normalized Root Mean Square Error (NRMSE) and correlation (R^2) are used for comparison.

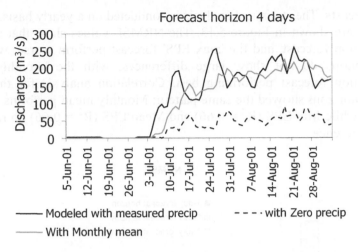

Figure 5.12 Example streamflow hindcasts with a 4-day forecast horizon. After 4 days the underestimation by assuming no rainfall becomes clear. Assuming Monthly mean rainfall shows a better comparison with the reference streamflow (simulated by using measured rainfall as input), but streamflow peaks, particularly in the beginning of the wet season, are underestimated.

Secondly flood warning verification is applied to compare forecasts of discharges above thresholds in terms of number of hits, missed events and false alarms (Van Andel et al., 2008[a]). Each of the measured discharge peaks that exceed the threshold is considered as one event. If the peak stays above the threshold for more than one day, this is still considered as only one event. If the forecasted discharge also exceeds the threshold, then the forecast is considered as hit. If the measured event is not forecasted, then it is called as missed event. The allowable time lag between the forecasted and measured events is taken as 2 days. The number of false alarms is the other important criterion used in warning verification. If the forecasted discharge exceeds the threshold when the measured discharge peak is below the threshold, then the forecast is considered as false alarm. If the forecast discharge stays above the threshold for more than one day, this is still considered only one false alarm. Such incidences can be identified with visual comparison of the measured and forecasted flows.

Visual inspection is the third important method of analysing forecast results. The interpretation of the statistical and flood warning verification results requires visual inspection.

5.5.5 Statistical verification

First the flow forecasts from each of the above discussed forecast methods are analysed by NRMSE. The simulated flow hydrograph from measured precipitation values is taken as reference when comparing with the different flow forecasts. The statistical analysis is conducted on a yearly basis. Results for 2000 are shown in Figure 5.13. The NRMSE values show that the 'zero precipitation forecast' and the 'max EPS' forecast perform worse, while the other options do not show large differences, with the 'monthly mean' precipitation forecast performing best. Correlation analysis for the 3-day forecast horizons showed the same pattern: Monthly mean performs best (R^2 = 0.79), while Min EPS (R^2 = 0.66) and Mean EPS (R^2 = 0.66) do not show much difference.

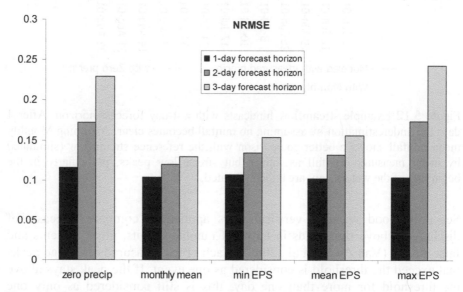

Figure 5.13 NRSME for 1, 2, and 3-day forecast horizons for the year 2000

It could be concluded from Figure 5.13 that the monthly mean, min EPS, and mean EPS forecasts result in better estimates than the other forecasts. However, this result alone cannot lead to a conclusion that these three forecasts are good estimates for flood forecasting, because statistical measures describe only general performance of models without special consideration of peak errors. Therefore, the monthly mean, min EPS, and mean EPS forecasts were further analysed by visual inspection.

5.5.6 Comparison by visual inspection

Forecasts with monthly mean precipitation and Min EPS forecasts as input underestimating peaks (Figure 5.14, Figure 5.15). The forecasts with Mean EPS precipitation as input are better catching the peaks (Figure 5.16). This shows that while Monthly mean forecasts showed the best performance with correlation analysis, the Mean EPS forecasts perform better with visual inspection. Capturing the peaks is of course crucial in flood forecasting applications. Therefore, the Mean EPS were analysed further for their applicability in flood forecasting and warning.

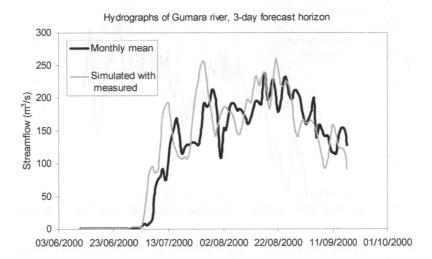

Figure 5.14. Gumara streamflow forecasts with Monthly mean precipitation forecasts as input to HEC-HMS, 3-day forecast horizon, wet season 2000

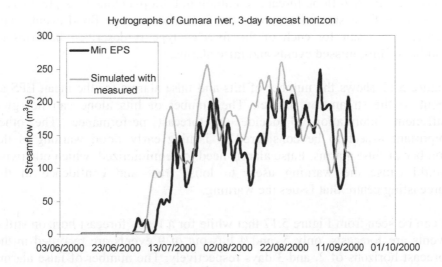

Figure 5.15. Gumara streamflow forecasts with Min EPS precipitation forecasts as input to HEC-HMS, 3-day forecast horizon, wet season 2000

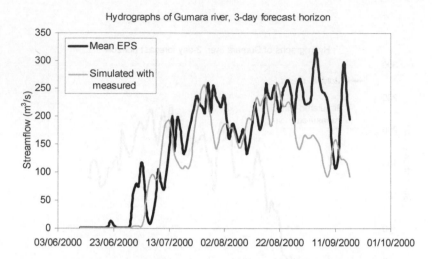

Figure 5.16. Gumara streamflow forecasts with Mean EPS precipitation forecasts as input to HEC-HMS, 3-day forecast horizon, wet season 2000

5.5.7 Flood early warning verification

The warning verification analysis is conducted on seasonal basis. The typical rainy season when flood threat is common in Ethiopia (June to September) is considered for a streamflow threshold indicated average flood events. The verification result for each of the forecast type is described in terms of number of hits, missed events and false alarms.

Figure 5.17 shows the number of hits and false alarms for the mean EPS as input to the streamflow model. The number of hits alone can not give sufficient information to decide on forecast performance. The other important aspect to be considered in issuing early flood warning is the number of false alarms. False alarms need to be minimized, which otherwise would cause the warning users to loose trust and confidence in the forecasting centre that issues the warning.

It can be seen from Figure 5.17 that while for a 1-day forecast horizon still 8 events are forecasts, only 5, and 4 hits out of 9 events are recorded in the forecast horizons of 2, and 3 days respectively. The number of false alarms is 11 for 3-day forecast horizon. The number of false alarms drops down for forecast horizons greater than 5-days. This shows that these forecasts do not anymore predict peak flows for more than the 5-days forecast horizon. Figure 5.17 shows clear decrease of forecast skill with increasing lead-time. The number of hits of 5 out of a total of 9 events for the 2-day forecast horizon is not very high. Again, visual inspection of the reference and forecasted hydrographs may clarify the patterns in the verification results.

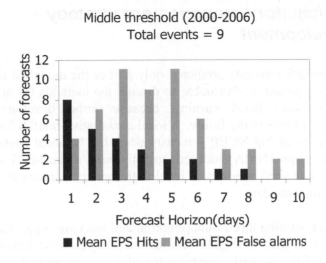

Figure 5.17 Number of hits and false alarms with mean EPS

Figure 5.18 shows that in 2002 the 2-day forecasts with mean EPS precipitation as input show good resemblance with the reference streamflow time series. The chosen warning threshold of 250 m^3/s leads to warnings for the peak events, however the timing should be further improved (there seems to be a delay in the forecasts). Application of the lower warning threshold of 210 m^3/s would increase the number of identified flood events, and the number of hits by the forecasts. Another way of increasing the number of hits is to look to ensemble forecasts between the Mean EPS and the Max EPS.

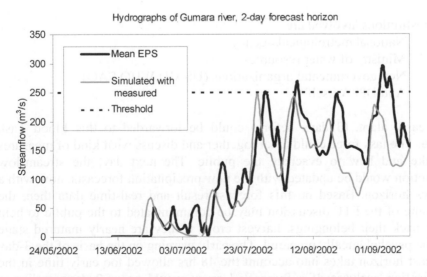

Figure 5.18 Mean EPS based flood forecast (2002, 2-days forecast horizon)

5.6 *Anticipatory management strategy development*

Although the EPS forecasts predicted only part of the modelled flood peaks in the analysis period it is valuable to discuss the methods for applying the EPS in operational flood warning, because further developments may improve the forecasts in the future. A local quantitative rainfall forecast may be nested in the global NCEP forecasts when the National Meteorological Agency of Ethiopia (NMA) succeeds to fully build up an MM5 model. The agency is now in semi-operational level with some constraints in initializing the model with local data.

Two different warning thresholds (medium and low) are suggested based on the results of the EPS forecasts verification analyses. The lower threshold could be used for an early warning for alert of operational services and decision makers, while the medium threshold could be used for issuing flood warnings. The verification analyses results suggest that the mean-EPS as input to the rainfall-runoff model provides the best predicitions.

The following two questions need to be addressed in order to issue an effective warning.
-    Who should first receive the warning and
-    When should the public receive it?

Following the countrywide severe flooding in 2006, a Flood Task Force (FTF) was set up in Ethiopia under the coordination of the Disaster Prevention and Preparedness Agency (DPPA).

The institutions involved are:
-    National metrological agency
-    Ministry of water resources
-    Non governmental organizations (USAID,WHO,FAO)
-    And DPPA itself

The early alarm (3 days before) could be forwarded to this Flood Task Force. The task force could meet together and discuss what kind of measures to take and how to evacuate the public. The next day, the streamflow prediction would be updated with the new precipitation forecasts, now with a 2-days horizon. Based on this forecast result and real-time data then, the outcome of the FTF discussion may be communicated to the public to help them pack their belongings, harvest crops if they are nearly matured stage and be psychologically prepared. The early warning on the basis of the 3-day forecast horizon takes into account the 48 hrs allowed too early time in the verification analyses. If a forecasted event would occur 2 days earlier as predicted, still mitigation measures would have been started.

5.7 Adoption of AWM in operational management policy

The results provide direction for further EPS research. The result guide to further research on EPS by considering ensemble forecasts between mean and max EPS using percentiles for probability thresholds as in the Rijnland case study (Chapter 4). Secondly, a short analysis of the forecasts showed that many of the flood forecasts are too late. This tendency of too late forecast has to be improved in either the rainfall forecast or the HEC-HMS model.

Both the HEC-HMS model and the ensemble precipitation forecasts need to be further improved before continuing with development of AWM strategies. The performance of the forecasting system is not good enough to consider application in the present form. Calibration and downscaling of the NCEP-EPS precipitation hindcasts, or replacement with the available higher resolution NCEP-GFS ensembles, are the preferred first steps to try and improve the forecasts. Calibration and statistical downscaling and analogues methods are unlikely to be effective, because of the limited number of monitoring stations in the area. Expanding the analysis area for the weather forecasts and complementing ground station data with remote sensing and re-analysis data can be used to overcome this problem. Dynamic downscaling will be possible by making use of the limited area MM5 model.

6 Conclusions and recommendations

6.1 Contributions to Anticipatory Water Management

Anticipatory Water Management (AWM) is defined as daily operational water management that pro-actively takes into account expected future conditions and events on the basis of weather forecasts. Anticipatory Water Management is an efficient way to optimise further the operational use of our water systems.

An approach to the development of Anticipatory Water Management strategies has been presented. This approach makes use of recent developments in weather forecasting, ensemble forecasting (providing forecasts of the dynamic probability distribution of the target variables) and water system control modelling. Flexible water system control models allow a wide range of control strategies to be applied in multi-year hindcast analysis. As archives of weather forecasts and water system state variables increase, hindcast verification analysis will become the basis for the development and optimisation of new control strategies.

Threshold based decision rules for early warning of critical events, on the basis of ECMWF EPS rainfall forecasts and hydrological simulation, have, for the first time, been verified and optimised for a hindcast archive of multiple years for a regional water system in the Netherlands. This is a valuable contribution, because increasingly water boards in the Netherlands are using ECMWF EPS rainfall forecasts for operational decision support.

Freely available NCEP EPS rainfall forecasts (both real-time and archived) and HEC-HMS hydrological simulation software were used to generate ensemble streamflow forecasts for a sub-catchment of Lake Tana in the Blue Nile basin, Ethiopia. This contribution shows that today much information for water management is freely available through the Internet and that the previously prohibitive costs and the lack of infrastructure for application of hydroinformatics decision support tools in less privileged countries is disappearing.

A method has been described to perform local, long-term verification analyses that are customized in order to evaluate probabilistic weather forecast products and to help in choosing the probability-threshold based decision rules for application in water management. Verification analysis methods from meteorology have been used and adapted. In meteorology, verification is done in terms of right or wrong decisions, e.g. hits, false alarms and misses. A suggested modification to the verification analyses applied in meteorology, is to verify on an event basis, instead of a fixed time step (e.g. daily).

Another step in the development of AWM that has been emphasised, is the simulation of controlled water systems. The modelling of a controlled water system is tested particularly by the high degree of freedom in both the system and the model, because of the control structures. The right output for the wrong reasons is a risk in using water system control models. On the other hand, many systems are (partly) manually operated or at least supervised by an operational water manager. The decisions of these managers are not as rigid as a computer simulated control strategy. Therefore getting a very close fit with a water system control model is mostly not possible.

A modelling approach has been formulated that takes advantage of the availability of a large amount of measurement data in controlled water systems. Water level and flow data at control structures allow for intensive validation and sub-system calibration to reduce the degree of modelling freedom, and to model separately the natural rainfall-runoff and hydro-dynamic processes. The remaining, unexplained, phenomena, which could not be captured by physically based modelling, are to be simulated with data driven modelling. The modelling approach has been applied to the Rijnland water system model, and has resulted in a clear improvement of the model as compared to a straightforward calibration and validation. It has been shown that by improving the long term volume balances of the model, also the short term water level simulations could be further improved. The resulting water system control model is more reliable for both design studies and operational decision support.

For integrated evaluation of AWM strategies, end-users (water boards) need to define their own criteria upfront. It has been shown for the Rijnland case study how these criteria can be expressed in a cost function. Then this cost function can be coupled to continuous simulation runs with the water system control model to analyse the dynamic cost-benefit analysis over a long period. This allows a search for least-cost alternatives and further optimisation of Anticipatory Water Management. It has been shown for the Rijnland case study that rule-based AWM strategies can be optimised using the water system simulation model with a Genetic search Algorithm (NSGAII). The presentation of a range of strategies along a Pareto Front, allows water managers to relate values of decision variables to requirements for different objectives, e.g. reducing flood or drought damage costs.

The process of developing AWM has been analysed and the identified steps have been cast in a framework. The application of this framework to the case studies prompts a review of the hypotheses posed at the beginning of this dissertation (Section 2.6).

6.2 Discussion of the hypotheses

Performance of hydro-meteorological ensemble forecasts over a long period of time for a particular catchment has been assessed by verification analysis with continuous simulation. The verification analysis of the ensemble precipitation and water level forecasts for the Rijnland case study confirms that:

> The comparison of measured precipitation and water level local to a given water system, with hydro-meteorological ensemble forecasts leads to an improvement in the use of those forecasts (hypotheses 1 and 2).

The verification, on the basis of decision rules for early warning for the need of anticipation, for different event thresholds showed that the number of false alarms decreased when higher thresholds were applied. The need for applying low probability thresholds and the high hit rate for forecast horizons from 5-8 days could also not have been known without the verification analysis. This is confirmed by the deviating warning rules as currently applied by the water board for the heuristic anticipation rules for their water system. Also, the case study of the Blue Nile shows that without verification analysis of the ensemble streamflow forecasts and warning thresholds, flood early warnings run the risk of missing all the flood events.

With all of the critical events of the Rijnland case study forecasted it can be concluded that:

> With hindcast analysis effective decision rules for early warning of the need for anticipation could be found (hypothesis 3).

With a maximum hit rate of 60% with respect to simulated reference streamflow, for the Blue Nile case study it is yet pre-mature to conclude the effectiveness of the forecasting system applied. Downscaling and calibration of the NCEP-EPS precipitation hindcasts, or replacement with the available higher resolution NCEP-GFS ensembles, and improvements of the rainfall-runoff models can further improve the effectiveness of the Blue Nile ensemble forecasts.

The long-term, continuous, simulation of the complete
AWM strategy for historic time series has enabled an
optimisation of AWM (hypothesis 4) for the Rijnland
case study.

A Pareto front of least flood damage or least drought damage cost (damage
costs of too high and too low water levels) using AWM strategies showed
clear convergence to optimal decision variables defining the dynamic switch
to AWM and the start and extent of the Anticipatory Water Management
action.

Unless the water authorities are forced to reduce the
flood damage costs, regardless of the costs of adverse
effects, a dynamic cost analysis, as applied in the
Rijnland case study, is needed to support the water
authorities in the decision whether or not to adopt
AWM (hypothesis 5).

Simpler cost-benefit analysis, e.g. with cost-loss ratio's, are not sufficient
because AWM does not concern a yes/no decision with constant cost-loss
ratio. Every event will be different from its predecessors. Note that even the
dynamic, continuous, cost analysis serves as a decision support analysis, not
as a decision model. It may be a strong argument in the decision process, but
incompleteness of the cost-model, uncertainties of the occurrence of critical
events and the performance of the forecasting system in the future, and
institutional, social and political arguments will be taken into account by
decision makers.

The main purpose of the cost-benefit analysis is to benchmark operational
water management strategies to assess whether the current strategy can be
improved and which alternative strategy is most efficient in doing so. As
such, the cost-benefit analysis with the optimisation approach can be used to
assess the current potential of AWM. For the Rijnland case study it can be
concluded that:

The use of ensemble precipitation forecasts to decide on
anticipatory management actions, in preference to re-
active management, can reduce the damage over a long
period of time (hypothesis 6).

A rule based, two-stage lowering of reservoir levels for flood control, on the
basis of warnings from ensemble water level forecasts, was shown to be
effective in reducing the estimated total costs of too high and too low water
levels, over an analysis period of 8 years, with 30% (Section 4.8.2).

Based on the AWM strategies found for the Rijnland case study, which reduce damage costs of too low and too high water levels, it is likely that the benefits when applying AWM, more than compensate for the losses when AWM is not applied (hypothesis 7). However, more research into the uncertainties of the expected benefits and losses is needed to confirm this hypothesis.

One source of uncertainty is how the analysed performance of the forecasting system for previous years corresponds to the performance of the coming years. Remaining research questions in this respect are discussed in the section on recommendations for further research (Section 6.5).

6.3 Conclusions

Anticipatory Water Management outperforms re-active operational water management or management on the basis of hydrological predictions alone.

ECMWF ensemble precipitation forecasts contain valuable information for anticipatory water management of regional water systems in the Netherlands. Hindcast verification analyses for the Rijnland water systems show that these forecasts are effective in reducing flood damage. The skill of the forecasts is such that the estimated reduction of flood damage is more than the increase in damage due to false alarms. Therefore, the use of ECMWF EPS in Anticipatory Water Management strongly reduces the total damage costs.

Freely available forecasting products, such as NCEP GFS, and hydrological simulation modelling systems, such as HEC-HMS, can be used to develop Anticipatory Water Management strategies world-wide at a low cost-level.

Given the variation in the ratio between the costs of false alarms and missed events on the one hand, and the benefits of hits and correct rejections on the other, a continuous cost model including the multi-objectives is the preferred evaluation criterion. Because of the complex, non-linear relationships between forecast, interpretation, management action, water system state, and long-term cost, evolutionary search algorithms are the preferred tool to expose the pay-offs between different AWM strategies and to enable the water authority to choose their optimal strategy.

Forecast archives are limited and contain only a few relevant events. Re-forecasting with new or updated meteorological products has to be performed for many previous years and including many relevant events. This is crucial to enable water authorities to develop anticipatory water management strategies and evaluate with confidence whether the water authority would benefit from applying AWM.

Anticipatory Water Management Strategies should be developed and optimised using continuous hindcast simulation to verify the accuracy of the forecast local to the water system of interest, and any decision rule as defined by the water authority.

The level of today's hydroinformatics tools to simulate off-line the complete process of real-time Anticipatory Water Management is such that continuous hindcast simulations for multi-year periods can be executed in a limited amount of time, with a level of realism that builds confidence, and with a degree of flexibility in defining decision rules that permits experimentation with different strategies.

6.4 Recommendations for management practice

The framework for developing Anticipatory Water Management (Section 3.9) is recommended for use as a process guideline to evaluate current operational water management strategies, and to improve these strategies with enhanced application of weather forecasts. Through experience with the case studies it is shown that the Anticipatory Water Management framework helps water managers and water management policy makers answer questions on how to develop, evaluate and decide on the adoption of AWM.

In this research the current operational management strategy of the Principal Water-board of Rijnland has been benchmarked with rule-based AWM strategies. The rule-based type of strategy matches with current operational practice. Not only the water board of Rijnland, but an increasing number of other water boards in the Netherlands, are working with rule-based early warnings using ensemble forecasts from the ECMWF. These rules are being set up on the basis of expert judgement and adjusted on a trial-and-error principle. There is an urgent need to verify these rules using hindcast analysis. For Rijnland, for example, the optimal decision rules found are different from the currently applied heuristic rules.

The ensemble water level forecasts for the Rijnland case study are currently running in real-time, in parallel to the operational DSS of Rijnland, to enable evaluation of the forecasting system by the operational water managers.

Re-analysis, hindcasting, and verification as facilitated by modelling systems are the vehicles for bridging the gap between theory and practice. In order to increase the application of weather forecast products meteorological organisations should provide more hindcast data sets to allow the end-users to assess the performance of the product for their intended use. For water authorities, in turn, it is essential for them to build up their water-system data and forecast archives.

Meteorological organisations are mostly limited in staff and computational power resources to perform hindcasting, while not risking interference with their operational tasks. Therefore, the responsibility of running and providing the hindcasts should be separated from the meteorological organisations that have operational forecasting responsibilities. Independend hindcasting institutes should be established. A funding model needs to be found with support from national governments and a wide variety of end-user groups.

A cost model of inappropriate and appropriate anticipatory management actions, and an adequate simulation model of the controlled water system are key in applying the analyses as described in this dissertation. These two requirements are often not readily available for a particular water system, or with a particular water authority. Development of these cost-models and simulation models of controlled water systems is recommended.

Space and time variability in predictions of the atmosphere and the water systems, is such that water authorities cannot rely on general performance indicators of the weather forcasts as provided by the meteorological institutes. Water authorities themselves should apply hindcast analyses local to the water system they are responsible for. The increasing availability of data and forecast archives, and simple-to-use re-forecasting and simulation technologies, have taken away the former practical and economic restrictions in doing this.

6.5 Recommendations for further research

Expanded set of decision rules
This research focussed on development, verification and optimisation of AWM strategies based on heuristic decision rules with probability threshold to capitalise on the ensemble forecasts. The choice was made to study the possibilities of training the decision rules on readily available forecasting products in the case study areas. The range of heuristic decision rules applied is not exhaustive. Further research on strategies with additional heuristic rules to find other important decision variables is recommended. One example of a potential additional decision variable is 'consistency'. This variable would support a decision for AWM actions if a certain number of subsequent forecasts indicate an upcoming critical event, whereas AWM would not be advised if only one of the subsequent forecasts indicates a critical event.

Anticipatory flood control with drainage canals

The anticipatory actions covered in the Rijnland case study concerned a temporary storage increase in the main discharge canals that make up the storage basin. In addition, the water level could be lowered in all the small drainage canals or ditches in the low-lying areas (polders) in anticipation of a critically excessive rainfall event to create extra storage. The potential of the anticipatory control for the drainage canals is larger than for the storage basin (about 60% more storage volume). Following the lower drainage level, also the groundwater level will drop, which again creates substantial extra storage in the soil (Schultz, 1992, p. 136-137).

Research into the anticipatory control of polder drainage networks will become relevant due to the current steps being taken to link the monitoring networks in the polders and the storage basins. The many small pumping stations that control the drainage network (200 in the Rijnland area), compared with 4 pumping stations for the storage basin in the Rijnland area, have to be controlled centrally and automatically, while presently most small pumping stations are still operated in local automatic or manual control modes. Research will have to be performed to check the controllability of the polder drains, because with their limited dimensions their discharge capacity might be the limiting factor instead of the discharge capacity of the pumping stations. The sensitivity of the land use (agricultural) and soil subsidence to the temporary groundwater level changes due to the anticipatory lowering of water levels in drainage canals has to be analysed. This again will result in particular requirements on the accuracy of the forecasting system in terms of hits and false alarms.

Anticipatory Water Management compared with structural measures

While this research has shown that the application of AWM will both reduce the frequency of flooding and damage costs of deviations from target water levels in the Rijnland water system, it will be even more beneficial to the water authorities if AWM could replace structural measures such as expanding the storage or discharge capacity.

In the Netherlands the required storage and discharge capacity of the channelled storage basins in regional water systems is determined on the basis of estimated return periods of exceedance of a critical water level. The required return periods depend on the potential flood damage in the catchment (risk-based approach; IPO, 2006). For example, for most parts of the channelled storage basin of Rijnland the return period of exceedance of the critical water level for flood damage should be 100 years or longer. The return period is estimated using historic extreme rainfall events applied to a water system simulation model. The design of structural measures to increase the water system's capacity to meet the return period requirement is done assuming no failure of these measures, e.g. pumping stations or the use of emergency storage basins. Indeed, the operational reliability of pumping stations (with adequate back-up systems in place) and, more disputable, of emergency storage basins, allows this assumption to be made.

The reliability of AWM in reducing flood frequency, compared with the reliability of a structural measure is less, simply because some events are missed by the forecasting system. The extra storage capacity made available at the beginning of an event by anticipatory pumping cannot be guaranteed. Therefore, in the first place, AWM should be seen as an optimisation of the use of an existing system, not as a (structural) change to the system. Secondly, AWM will increasingly need to be applied due to societal demands. This is because weather forecasts are becoming available to professional and public stakeholders, who increasingly take note of the information. Then, for example, if flood damage occurs, and forecasts of extreme rainfall have been given, water authorities will be asked to justify their decisions if anticipatory actions were not taken. There will be growing demands to the water authorities to use all the information available.

Assessing the effectiveness of AWM
Although AWM should not be viewed as a structural measure to increase the capacity of a water system, when applied it will reduce the frequency of system failure (e.g. exceedance of critically high water levels) and thus reduce the need for a structural increase in the system capacity. The main research question then becomes: How much will AWM reduce the frequency of failure? Therefore, the use of AWM should be taken into account when estimating the return periods of failure. How to do this is an important and challenging topic for further research.

Importantly, the extensive archive of recorded extreme rainfall events used for the frequency analysis is generally not accompanied by weather forecasts for the same events, which are needed to simulate AWM. Whether and how these large archives of coupled extreme events and probabilistic weather forecasts can be created still needs to be determined. How many critical events and forecasts are needed to determine the effectiveness of AWM with sufficient statistical reliability is another question to be answered. If, in addition to the effectiveness, also the (economic) efficiency is to be determined, then also the statistics of normal hydro-meteorological conditions and their forecasts (false alarms or correct rejections) need to be assessed.

Risk-based AWM versus Rule-based AWM
In a follow-up of this research, the risk-based AWM strategy will also be verified and benchmarked for the same data set. It is hypothesised that the rule-based AWM strategy might be more successful than the risk-based approach. This is because the risk-based approach assumes perfect probabilistic predictions by minimising the expected damage for every time-step. The ensemble hydro-meteorological forecasts are not perfect

probabilistic predictions. However, probabilistic hydro-meteorological forecasts are continuously improving and pre- and post-processing techniques (e.g. downscaling and bias correction) can be used to fine-tune the forecasts local to the case study area.

Establishing the level of quality of the probabilistic forecasts for which risk-based AWM becomes more cost efficient than rule-based AWM would be interesting additional research. A second limitation of the (minimum) risk-based approach is that it is expected to be highly governed by the ensemble average, which may result in too little anticipation to reduce damage of extreme events considerably. Weights can be used in the objective functions to take preferences for risk-averse decisions into account, but as soon as weight factors are introduced, the strategy moves towards strategies of heuristic rules, which need similar optimisation approaches as described in this dissertation.

Next to the arguments described above, there is another reason why the rule-based AWM approach may be preferred from an operational water management point of view. This is because risk-based approaches, with risk defined as probability of occurrence times damage, inherently provide only an expected cost over a long period of time. In day-to-day decisions, from event-to-event, it is not 'Risk' that matters. The water board will not be confronted by the average, expected damage, but always with either the maximum, or the minimum damage as a consequence of the momentary decision. Operational managers want to be aware of the maximum damage that may occur. Research into the comparison and combination of minimum risk and rule-based AWM strategies is recommended. The most favourable strategy will be somewhere in between, differently for every case study, depending on the quality of the probabilistic forecast, and the requirements of the water authority.

AWM and climate change
AWM permits a more flexible use of water systems to optimise the management of critical and extreme events. Because it is based on real-time meteorological forecasts it increases the preparedness and adaptivity to climate change. At the same time climate change adds to the uncertainty of the performance of weather forecasts in the future. This cannot be an excuse not to use all the currently available information as effectively as possible in managing our water systems.

However, we should research and monitor the potential challenges for AWM in a changing climate. For example, statistics of critical events and forecast accuracy may change. This means that optimal decision rules may also change and that statistical downscaling and bias correction methods may fail. Because atmospheric simulation models are physically based, and because of the use of monitoring data for the initial state, it may be assumed that for a large part forecast accuracy is independent of climate change. Frequent re-calibration of the models would further contribute to maintaining and improving forecast skill in a changing climate. Research to enhance methods of calibration and optimisation methods that accomodate sudden changes in trends remain of the upmost value in this respect.

While the development of numerical weather prediction so far has shown improvement or at least the maintenance of weather forecast performance, it is not inconceivable that for some catchments, in case of sudden climate change, the frequency of events that are difficult to predict increases faster than the numerical weather prediction can keep up with (for example occurrence of convective rainfall). Such (temporal) decrease of forecast accuracy should be signalled quickly and an adjustment of the AWM strategy should be considered. Research that verifies assumptions about the behaviour of the peformance of weather forecasts in a changing climate is needed.

Wider applicability of AWM

This research focussed on AWM for flood forecasting, early warning, control and evacuation applications through the two case studies. The potential for wider application was illustrated in Chapter 1, Figure 1.1. Applications identified there were, amongst others, hydropower, water supply, irrigation and urban drainage. These applications are expressed in terms of the *end-use* of the water system, but often the requirements for these different end-uses need to be met simultaneously for the same water system. Therefore, the applicability of AWM can also be described in more general terms. The need and effectiveness of AWM depends for each case study on the *spatial scale* of the water system, the requirement for the *forecast range*, the type of *hydrological problem*, and the *controlability* of the water system. These characteristics will determine the accuracy of the hydro-meteorological forecasts available and the potential effect of the anticipatory actions. The (economic) efficiency in addition depends on the *benefit* from effectively anticipated events and the *adverse effects* of false alarms and missed events. The benefits and adverse effects should be compared to current practice in which damage occurs as well, due to taking actions late or taking no control actions at all.

In general the larger the spatial scale of the water system, the lower the resolution requirements for the meteorological forecasts. With regard to the forecast range, monthly and seasonal forecasts of precipitation and potential evaporation, for example, are applied to reservoir control to assure water supply throughout the year. It becomes clear that there is a potential to use weather forecasts for all hydro-meteorological variables for all forecast ranges, from nowcasting to short range (up to 2 days) and from medium range (2-10 days) to long range (monthly and seasonal).

The hydrological problems for which AWM is needed can be grouped in traditional problem descriptions for water management as 'too much water', 'too little water', and 'poor water quality'. While the flood control applications concern problems with too much water, AWM can also be applied for problems with too little water (e.g. decisions for water inlets during dry spells can be taken on the basis of rainfall forecasts for the coming days) and for problems with water quality (e.g. control can be optimised to minimise CSO's from urban systems on the basis of now-casts and short term rainfall forecasts).

The controlability of the water system determines the type and effectiviness of anticipatory actions. For example, the controlability for the Rijnland case study is limited because of the strong adverse effects of too low water levels. Other regional water systems in the Netherlands can lower the water levels further with less adverse effects, and vice versa with less controlability and more adverse effects.

We will continue to research the applicability of AWM to case studies covering the full range of characteristics described above. Scientists, engineers and practitioners are called on to join in an effort to maximise the use of hydro-meteorological forecasts in operational water management.

References

Abbott, M.B., 1991: Hydroinformatics: Information Technology and the Aquatic Environment. Ashgate, Aldershot, UK, and Brookfield, USA.

Abbott M.B., Refsgaard J.C., 1996: Distributed Hydrological Modelling, Kluwer Academic Publishers, Dordrecht, The Netherlands, pp.321

Abbott, M.B., 1999: Introducing Hydroinformatics, Journal of Hydroinformatics, Vol. 1, No. 1, 3-20

Abbott, M.B, 2005: The waterknowledge initiative, Knowledge Engineering and European Institute for Industrial Leadership, Brussels

Andel, S.J. van, Price, R.K., Lobbrecht, A.H., Kruiningen, F. van, Mureau, R., 2008[a]: Ensemble Precipitation and Water-Level Forecasts for Anticipatory Water-System Control, J. Hydrometeor., 9, 776–788

Andel, S.J. van, Lobbrecht, A.H., Price, R.K., 2008[b]: Rijnland case study: hindcast experiment for anticipatory water-system control, Atmospheric Science Letters, Vol. 9, No 2, 57-60

Andel, S.J. van, Price, R.K., Lobbrecht, A.H., Kruiningen, F. van, 2009[a]: Modelling controlled water systems, J. of Irrigation and Drainage. *in press*

Andel, S.J. van, Price, R.K., Lobbrecht, A.H., Kruiningen, F. van, Mureau, R., 2009[b]: Framework for Anticipatory Water Management: application in flood control for Rijnland reservoir system, *submitted*

Akhtar, M. K., Corzo, G. A., van Andel, S. J., Jonoski, A., 2009: River flow forecasting with artificial neural networks using satellite observed precipitation pre-processed with flow length and travel time information: case study of the Ganges river basin, Hydrol. Earth Syst. Sci., 13, 1607-1618

Assefa, K.A., 2008: Flood forecasting and early warning in Lake Tana sub-Basin, Upper Blue Nile, Ethiopia, UNESCO-IHE MSc Thesis WSE-HI.08-08

Atger, F., 2001: Verification of intense precipitation forecasts from single models and ensemble prediction systems, Nonlinear Processes in Geophysics, Vol. 8, 401-417

Bálint, G., Csík, A., Bartha, P., Gauzer, B., 2005: Application of Meteorological Ensembles for Danube Flood Forecasting and Warning, NATO Advanced research Workshop Proceedings, Trans–boundary Floods: Reducing Risks and Enhancing Security through Improved Flood Management Planning, Oradea B.F. (ed.), 282-293, Oradea, Romania, Treira S.R.L.

Barreto, W., Price, R.K., Solomatine, D.P., Vojinovic, Z., 2006: Approaches to multi-objective multi-tier optimization in urban drainage planning, 7th International Conference on Hydroinformatics, 2006, Nice, FRANCE

Bashar, K.E., Zaki, A.F., 2006: SMA Based Continuous Hydrologic Simulation of The Blue Nile, Proceeding of the International Conference for the FRIEND/Nile Project Towards a Better Cooperation, Sharm El Shiekh, Egypt, 12-14 November 2005

Belaineh, G., Peralta, R.C., Hughes, T.C., 1999: Simulation/Optimization Modeling for Water Resources Management, Journal of Water Resources Planning and Management, 125(3), 154-161

Beven, K., Freer, J., 2001: Equifinality, data assimilation, and uncertainty estimation in mechanistic modelling of complex environmental systems using the GLUE methodology, Journal of Hydrology, Vol. 249, 11-29

Bhattacharya, B., Lobbrecht, A. H., Solomatine, D. P., 2003: Neural Networks and Reinforcement Learning in Control of Water Systems, Journal of Water Resources Planning and Management, Vol. 129, No. 6, 458-465

Boetzelaer, M. van, Schultz, B., 2005: Historical Development of Approaches and Standards for Flood Protection along the Netherlands Part of the Rhine River, ICID 21st European Regional Conference, 2005, Frankfurt (Oder) and Slubice, Germany and Poland

Bokhorst, J., Lobbrecht, A.H., 2005: Neerslag-kansverwachting voor het waterbeheer. (Probability precipitation forecasts for water management). H2O, 13, 26–29 (in Dutch)

Buizza, R., Houtekamer, P.L., Toth, Z., Pellerin, G., Wei, M., Zhu, Y., 2005: Comparison of the ECMWF, MSC, and NCEP Global Ensemble Prediction Systems. Mon. Wea. Rev., 133, 1076-1097

Carpenter, T.M., Sperfslage, J.A., Georgakakos, K.P., Sweeney, T., Fread, D.L., 1999: National threshold runoff estimation utilizing GIS in support of operational flash flood warning systems, J. Hydrol. 224(1), 21-44

Chu, X., Steinman, A., 2009: Event and Continuous Hydrologic Modeling with HEC-HMS, Journal of Irrigation and Drainage Engineering, 135(1), 119-124

Clark, M.P., Gangopadhyay, S., Hay, L., Rajagopalan, B., Wilby, R., 2004: The Schaake shuffle: A method for reconstructing space–time variability in forecasted precipitation and temperature fields, J. Hydrometeor., 5, 243-262

Clemmens, A.J., Bautista, E., Wahlin, B.T., Strand, R.J., 2005: Simulation of Automatic Canal Control Systems, Journal of Irrigation and Drainage Engineering, 131(4), 324-335

Coello, C.C.A., 2005: Twenty Years of Evolutionary Multi-Objective Optimization: A Historical View of the Field

Deb, K., Pratap, A., Agarwal, S., Meyarivan, T., 2002: A Fast and Elitist Multiobjective Genetic Algorithm: NSGA–II., IEEE Transactions on Evolutionary Computation, 6(2):182-197

DFO., 2008; Dartmouth Flood observatory www.dartmouth.edu/~floods/index, Accessed; March, 2008

DPPC, 2007: Woreda Agriculture and Rural Development Office

ECMWF, 2006: The Catalogue of ECMWF Real-Time Products, www.ecmwf.int/products/catalogue

ECMWF, 2007: The evolution of the ECMWF analysis and forecasting system. (www.ecmwf.int/products/data/operational_system/evolution)

Falkovich, A., Kalnay, E., Lord, S., Mathur, M. B., 2000: A New Method of Observed Rainfall Assimilation in Forecast Models, Journal of Applied Meteorology: Vol. 39, No. 8, 1282-1298

Ferraris, L., Gabellani, S., Rebora. N., 2003: A comparison of stochastic models for patial rainfall downscaling, Water Resources Research, VOL. 39, NO. 12, 1368, SWC 12

Franz, K., N. Ajami, J. Schaake, R. Buizza, 2005: Hydrologic Ensemble Prediction Experiment Focuses on Reliable Forecasts, Eos, Trans. Amer. Geophys. Union, 86(25), 239

Hamill, T., 2009: An overview of the use of reforecasts for precipitation forecast calibration, HEPEX workshop, 15-18 June, 2009, Toulouse

HKV LIJN IN WATER, 2006: Toekomstig Waterbezwaar Rijnland,

Hoofdrapport fase 2 en 3, Project report, PR981.10, 59 pp. (in Dutch)

Hlavcova, K., Szolgay, J., Kubes, R., Kohnova, S., Zvolenský, M., 2005: Routing of Numerical Weather Prediction System Outputs through a Rainfall–Runoff Model. NATO Advanced research Workshop Proceedings, Trans–boundary Floods: Reducing Risks and Enhancing Security through Improved Flood Management Planning, Oradea B.F. (ed.), 105–116, Oradea, Romania, Treira S.R.L..

Hoes, O.A.C., 2007: Aanpak wateroverlast in polders op basis van risicobeheer, Optima Grafische Communicatie, Rotterdam, ISBN 978-90-9021597-6, 188 pp. (in Dutch)

Holly, F.M. Jr., Merkley, G.P., 1993: Unique Problems in Modeling Irrigation Canals, Journal of Irrigation and Drainage Engineering, 119(4): 656-662

Huang, W.-C., Yuan, L.-C., 2004: A drought early warning system on real-time multireservoir operations, Water Resources Research, Vol. 40, No. 6., W06401

HydroNet, 2009: Software, www.hydronet.nl

IPO, 2006: IPO-richtlijn ter bepaling van het veiligheidsniveau van boezemkaden (Guideline for determining safety standards of embankments of regional storage basin channels), (in Dutch)

Kebede, S., 2006: Water balance of Lake Tana and its sensitivity to fluctuations in rainfall, Blue Nile basin, Ethiopia, Journal of Hydrology 316, Issues 1-4, 233-247

Kok, C.J., 2000: On the behaviour of a few popular verification scores in yes/no forecasting, KNMI Scientific Report, WR-2000-04, De Bilt, The Netherlands, 73 pp.

Krzysztofowicz, R., 2001: The case for probabilistic forecasting in hydrology. J. Hydrol., 249, 2-9

Krzysztofowicz, R., 2002: Bayesian system for probabilistic river stage forecasting, Journal of Hydrology, Vol. 268, 16-40

Kummerow, C., 2000: The status of the Tropical Rainfall Measuring Mission (TRMM) after two years in orbit, J. Appl. Meteor., 39, 1965-1982

Legg, T., Mylne, K., 2003: Numerical Weather Prediction Severe weather early warnings from ensemble forecast information, Forecasting Research Technical Report No. 415, Met Office, Bracknell, UK

Lobbrecht, A.H., 1997: Dynamic Water-System Control, Design and Operation of Regional Water-Resources Systems, Dissertation, A.A. Balkema Publishers, ISBN 9054104325

Lobbrecht, A.H., Dibike, Y.B., Solomatine, D.P., 2002: Applications of Neural Networks and Fuzzy Logic to Integrated Water Management, STOWA Project Report, Delft, the Netherlands

Lobbrecht, A.H., Hiemstra, G., Talsma, M., Vonk, Z., 2003: Neerslaginformatie voor het waterbeheer, H2O, Vol. 23 (in Dutch)

Lobbrecht, A.H., Loos, S., 2004: From Precipitation Data to Flood Forecasting, 6th International Conference on Hydroinformatics - Liong, Phoon & Babovic (eds), World Scientific Publishing Company

Lobbrecht, A.H., Solomatine, D.P., 1999: Control of water levels in polder areas using neural networks and fuzzy adaptive systems, Water Industry Systems: Modelling and Optimization Applications, Vol.1, D. Savic, G. Walters (eds.). Research Studies Press, Ltd. Baldock, UK,. 509-518

Lutgens, F. K., Tarbuck, E. J., 2001: The Atmosphere, An Introduction to Meteorology, 8th edition, Prentice Hall, ISBN 0130879576, New Jersey, US

Maidment, D.R., 1994: Hand book of Hydrology

Maskey, S., 2004: Modelling Uncertainty in Flood Forecasting Systems, Dissertation, ISBN: 978-90-5-809694-4

Maskey, S., Andel, S.J. van, Venneker, R., Uhlenbrook, S., 2008: A fuzzy-probabilistic approach to support flood warning decision making, *submitted*

Moll, J.R., Parmet, B.W., Sprong, T.A., 1996: Country paper of the Netherlands, Proceedings of Management and Prevention of Crisis Situations: Floods, Droughts and Institutional Aspects. No. 2, EurAqua, European Network of Fresh Water Research Organizations: 123-142

Molteni, F., Buizza, R., Palmer, T.N., Petroliagis, T., 1996: The ECMWF Ensemble Prediction System: Methodology and validation. Quart. J. Roy. Meteor. Soc., 122, 73-119

Mureau, R., Molteni, F., Palmer, T.N., 1993: Ensemble Prediction using dynamically conditioned perturbations, Q.J.R. Meteorol. Soc., 119, pp. 299-323

NASA, 2008: TRMM data interface. http://disc2.nascom.nasa.gov/Giovanni/tovas/. Last visited: 7-10-2008

Nash, J. E., Sutcliffe, J. V. , 1970: River flow forecasting through conceptual models. Part 1: A discussion of principles, Journal of Hydrology, 10(3): 282-290

NCEP, 2008: Global Forecast System, ftp://ftp.cdc.noaa.gov/pub/Datasets.other/refcst/ensdata/

Nestler, J.M., Goodwin, R.A., Loucks, D.P., 2005: Coupling of engineering and biological models for ecosystem analysis, Journal of Water Resources Planning and Management, 131(2): 101-109

NHS, 2002: Information for Action, A good practice guide on anticipatory management in healthcare, Department of Health, Healthcare Operational Intelligence Project, http://www.dh.gov.uk/assetRoot/04/06/30/16/04063016.pdf

Olsson, J., Lindstrom, G., 2008: Evaluation and calibration of operational hydrological ensemble forecasts in Sweden, Journal of Hydrology, Volume 350, Issues 1-2, 14-24

Overloop, P.J. van, 2006: Model Predictive Control on Open Water Systems, Ph.D. dissertation, 2006

Park, J., Obeysekera, J., Zee, R. van, 2007: Multilayer Control Hierarchy for Water Management Decisions in Integrated Hydrologic Simulation Model, Journal of Water Resources Planning and Management, 133 (2): 117-125

Persson, A., Grazzini, F., 2007: User Guide to ECMWF forecast products, Meteorological Bulletin M3.2, ECMWF, UK

Price, R.K., 2006: Engineering and technical aspects of managing floods: hazard reduction, operational management and post-event recovery. Frontiers in Flood Research, Kovacs Colloquium, IAHS publ. 305, 73-91

Price, R.K., 2008: Hydroinformatics in retrospect, UNESCO-IHE, Delft, The Netherlands, ISBN 978-90-73445-20-8

Rayner, S., Lach, D., Ingram, H., 2005: Weather Forecasts are for Wimps: Why Water Resource Managers Do Not Use Climate Forecasts, Climatic Change, 69 (April 2005), 2–3, 197-227

Richardson, D.S., 2000: Skill and relative economic value of the ECMWF ensemble prediction system. Quart. J. Roy. Meteor. Soc., 126, 649-667

Rijnland, 2000: Studie Toekomstig Waterbezwaar. (Study of future hydrological loads). Project report, Rijnland Water Board, Leiden, the Netherlands (in Dutch)

Roo, A.P.J. de, and Co-authors, 2003: Development of a European flood forecasting system. Intl. J. River Basin Management, 1 (No. 1), 49-59

Roulin, E., 2007: Skill and relative economic value of medium–range hydrological ensemble predictions. Hydrol. Earth Syst. Sci., 11, 725-737

Roulin, E., Vannitsem, S., 2005: Skill and Medium–Range Hydrological Ensemble Predictions. J. of Hydrometeor., 6, 729-744

Sattler, K., Feddersen, H., 2003: Final report for the special project SPDKEFFS: Heavy rain in Europe, Danish Meteorological Institute, Copenhagen, Denmark

Schaake, J.C., Franz, K., Bradley, A., Buizza, R., 2006: The Hydrologic Ensemble Prediction EXperiment (HEPEX), Hydrol. Earth Syst. Sci. Discuss., 3, 3321-3332

Schaake, J.C., Hamill, T.M., Buizza, R., Clark, M., 2007: HEPEX: The Hydrological Ensemble Prediction Experiment, BAMS, 88(10), 1541-1547

Schilling, W., 1990: Foreword, in Applications of Operations Research to Real Time Control of Water Resources Systems, Proceedings of the First European Junior Scientist Workshop, Kastanienbaum/Luzern, Switzerland, 15-18 March, edited by Einfalt, T., Grottker, M., Schilling, W., Schriftenreihe der EAWAG, CH-8600, Dübendorf, Switzerland

Schultz, E., 1992: Waterbeheersing van de Nederlandse droogmakerijen, (Water management of the drained lakes in the Netherlands), Dissertation T.U. Delft, ISBN 90-369-1087-0

SMEC, 2006: Project preparation, Flood prepardness and early warning, Technical back ground paper, volume1: Main report. Addis Ababa, Ethiopia, ENTRO

Smith, K., Ward, R., 1998: Floods: Physical processes and human impacts, John Wiley & Sons, Chichester, England

SRTM, 2008: Digital Elevation Model data, global coverage, www.ambiotek.com/srtm. Last visited: February 2008

Swinkels, M.A.J., 2004: Control rules for operational management of polder storage basin systems, M.Sc. Thesis, Technical University Delft (in Dutch)

Thielen, J., Bartholmes, J., Ramos, M.-H., Roo, A. de, 2009: The European Flood Alert System – Part 1: Concept and development, Hydrol. Earth Syst. Sci., 13, 125-140

Todini, E., 1999: Using phase-state modelling for inferring forecasting uncertainty in nonlinear stochastic decision schemes, Journal of Hydroinformatics, 01.2, 75-82

USACE, 2003: Hydrological Modelling System HEC-HMS, Technical Manual, Version 3.1., US Army Corps of Engineers, Hydrological Engineering Center

Vriend, H.J. de, 2002: Model-based morphological prediction: from art to science, In: River Flow 2002, Eds. D. Bousmar, D.; Zech, Y.; Proc. Int. Conf. Fluvial Hydraulics, Louvain-la-Neuve, Belgium, 4-6 September 2002, 3-12

Wandee, P., 2005: Optimization of water management in polder areas : some examples for the temperate humid and the humid tropical zone, Dissertation, Taylor & Francis, London

WB21, 2000: Waterbeleid voor de 21e eeuw, Geef water de ruimte en de aandacht die het verdient, Advies van de Commissie Waterbeheer 21e eeuw, Water policy for the 21th century (in Dutch), pp 97

Weijs, S.V., Leeuwen, P.E.R.M. van, Overloop, P.J. van, Giesen, N. van de, 2007: Effect of uncertainties on the real time operation of a lowland water system in the Netherlands, IAHS Publ. 313, 463-470

Werner, M., Reggiani, P., Roo, A. de, Bates, P., Sprokkereef, E., 2005: Flood Forecasting and Warning at the River Basin and at the European Scale, Natural Hazards, 36, 25–42

WMO, 2004: Report of the CHy Working Group on Hydrological Forecasting and Prediction, Commision for Hydrology, CHy-XII/Doc. 6, 7.VII.2004

Yufeng, G., 2003: Flood Modelling and Forecasting for Rijnland Area Using Physically Based and Data-Driven Methods, M.Sc. Thesis, HH 470, UNESCO-IHE, Delft, the Netherlands

List of Figures

Figure 1.1 Significance of meteorological forecasts for operational water management applications 22

Figure 1.2 Basic representation of the process of Anticipatory Water Management 25

Figure 2.1 Elements of a hydro-meteorological ensemble prediction system 44

Figure 2.2 Elements of a hydro-meteorological ensemble prediction system expanded with a Decision support pre-processor for end-use of the predictions in Anticipatory Water Management 47

Figure 2.3 ECMWF EPS precipitation time series for location De Bilt (NL) (data source: KNMI). When applying threshold-based decision rules for EPS, the event threshold (Precipitation threshold), the forecast horizon and the probability threshold have to be set. The probability threshold is the required forecasted probability that the precipitation threshold will be exceeded. This is determined by the ensemble members exceeding the precipitation threshold. 49

Figure 3.1 Unnecessary high measured water levels (a) and high measured water levels prevented by early lowering of storage level (b) 56

Figure 3.2 Upper and lower precipitation thresholds for accumulated precipitation in Rijnland. After seven days (veritical line) the minimum threshold does not increase anymore. 59

Figure 3.3. Example of anticipatory action. Reservoir level is lowered in anticipation of a flood event. As a result of the anticipatory lowering, the resulting peak reservoir level is reduced. 60

Figure 3.4 General process of anticipatory water management 62

Figure 3.5 Fictitious example of a three-member ensemble precipitation forecast. At a certain lead-time the uncertainty might be considered too high for decision making (Tmax). 64

Figure 3.6 Creating hindcasts. The forecasting process is repeated for every time step t in the past. 65

Figure 3.7 Framework for modelling controlled water systems. Visualisation, discussion and modelling of the unknown processes are key. If the processes can be identified (e.g. by error analysis for different time scales) and isolated after visualisation and discussion, they can be represented by an internal or external physically based model, if not, a data driven approach can be used. 70

Figure 3.8 Fictitious example of a decision rule, based on three members of an ensemble hydro-meteorological forecast. 71

Figure 3.9 Risk based decision tree for flood warning, where the alternatives are W = {0, 1} and the future states of the system are F = {0, 1}. W = 0 and W = 1 imply "do not issue warning" and "issue warning", respectively. Similarly, F = 0 and F = 1 imply "the area is flooded" and "the area is not flooded", respectively. (Cited from Maskey et al., 2008) 72

Figure 3.10 Total cost estimation for alternative operational water management strategies 73

Figure 3.11 Pareto front for a 2-objective (criteria) optimisation problem with AWM strategies 78

Figure 3.12 Framework for developing Anticipatory Water Management. The main part of the framework consists of steps for screening of new forecast products and control strategies. If new control strategies perform well, in the next step the optimisation of the AWM strategy can be performed. 80

Figure 4.1 Principal Water-board of Rijnland: controlling a low-lying regional water system in the western part of the Netherlands. A channelled storage basin collects all the excess water of the area. The water level in the storage basin is controlled by four pumping stations. 82

Figure 4.2 Water level control of the Rijnland storage basin, with and without forecasting. When using forecasts and temporarily allowing lower water levels, extra storage of 2.2 x 106 m3 can be created before the extreme event occurs. 83

Figure 4.3 Comparison of radar and ground station precipitation estimates for the Rijnland area. The graphs show close resemblance for both dry and wet periods. 85

Figure 4.4 Aquarius water system control model of Rijnland (Yufeng, 2003) 87

Figure 4.5 Calibration of Aquarius water system control model of Rijnland, the Netherlands, for a peak water level event in November 2000. 91

Figure 4.6 Calibration of Aquarius water system control model of Rijnland, the Netherlands, for a normal flow period in February and March 2000. 91

Figure 4.7 Validation of the Aquarius water system control model of Rijnland, on the basis of monthly pumped discharge volumes. 91

Figure 4.8 Cumulative pump discharge volume from the Rijnland storage basin. Modelled volume is too low, because of underestimation during the dry summer seasons. 93

Figure 4.9 Cumulative pump discharge volume for events in the wet winter season in 2000(a) and 2001(b). For both events modelled volume is higher than measured volume, indicating over-estimation of the model during excess water events in the wet season. 93

Figure 4.10 Time scale analysis of difference between measured and modelled pump discharge. At 90-days moving average a clear sine function with a yearly period becomes visible. 95

Figure 4.11 Sine function to model the slow processes error (90-days moving average) of the Rijnland Aquarius model. 96

Figure 4.12 Monthly pumped discharge volume from Rijnland storage basin in 2002 of the final model. Note the improvements in summer months and October and November compared to Figure 4.7. Note also the accurately modelled total yearly volume (0.7% error), while only the total volume over 6-year simulation (1997-2002) was calibrated. 98

Figure 4.13 Cumulative pump discharge volume for events in the wet winter season in 2000(a) and 2001(b) after external modelling of unknown processes and calibration. 99

Figure 4.14 Calibration of the event of November 2000, after the unknown processes had been included as external data driven models. The modelling of the peak has improved considerably with respect to the first model (Figure 4.5) 99

Figure 4.15 Model results for a normal flow period in February 2000, after the unknown processes had been included as external data driven models. There are not many differences with the first model (Figure 4.6) 99

Figure 4.16 Nash-Sutcliffe coefficients for the Rijnland model, with the sine function and with a constant flow correction, for different time intervals. 100

Figure 4.17. Cumulative pump discharge volume from the Rijnland storage basin during calibration and validation. Note that the modelled cumulative volume now matches very well compared to the first model (Figure 4.8) and that the development of cumulative volume remains accurate during the validation period of 2003 and 2004. 101

Figure 4.18. Cumulative discharge volume for validation events in January and December 2003. 101

Figure 4.19 Contours of number of hits (a) and number of false alarms (b) of the ECMWF EPS precipitation forecasts for 85 precipitation events in the Rijnland water system of 15 mm day-1 or more. (ECMWF EPS precipitation for location De Bilt, from 25 April 1997 to 31 August 2004) 106

Figure 4.20 Detailed analyses of performance of threshold-based decision rules with ECMWF EPS precipitation forecasts. (a) Number of hits, events that have been forecasted too early, missed events and false alarms for a 15 mm day-1 precipitation threshold and a 3-day forecast horizon. (b) Measured daily precipitation and forecasted daily precipitation for a 3-day forecast horizon and a probability threshold of 0.04 (96th percentile). 107

Figure 4.21 Comparison of performance of decision rules based on precipitation forecasts and water level forecasts. [Left] Contours of number of hits (a) and false alarms (c) with ensemble precipitation forecasts for nine selected events. Winter precipitation threshold: 40 mm per 3 days. Summer precipitation threshold: 45 mm per 3 days. [Right] Contours of number of hits (b) and false alarms (d) with ensemble water level forecasts for nine selected events. Winter water level threshold: -0.57 m+Ref for 12 hours. Summer water level threshold: -0.55 m+Ref for 12 hours. 109

Figure 4.22 Comparison of performance of decision rules based on precipitation forecasts and water level forecasts. [Left] Contours of number of hits (a) and false alarms (c) with ensemble precipitation forecasts for nine selected events. Winter precipitation threshold: 65 mm per 5 days. Summer precipitation threshold: 65 mm per 5 days. [Right] Contours of number of hits (b) and false alarms (d) with ensemble water level forecasts for nine selected events. Winter water level threshold: -0.57 m+Ref for 12 hours. Summer water level threshold: -0.55 m+Ref for 12 hours. 110

Figure 4.23 Event based ROC-diagrams of ensemble precipitation (a) for 5, 6, and 7 days forecast horizons. For comparison the ROC curves of the 6-day forecast horizon water level forecasts and precipitation forecasts for 65 mm/5 days have been plotted (b). The curves show the relationship between hit rate and false alarm rate for different probability thresholds. The lowest probability threshold is the upper right end of the curves, for the highest probability thresholds the curves reach the origin (no hits and no false alarms). 112

Figure 4.24 Ensemble water level prediction of 11 September 1998 for the Rijnland water system on the basis of ECMWF EPS precipitation forecasts (Van Andel et al., 2008[b]). 115

Figure 4.25 Effect of modelled Anticipatory Water Management on water level control in the Rijnland water system. The modelled peak is lower than the measured peak, because in the model the water level is lowered before the precipitation event occurs. Exceedance of the -0.57 m+Ref level is prevented (horizontal line). 116

Figure 4.26 Water level-cost function Rijnland storage basin (Van Andel et al., 2009[b]). 118

Figure 4.27 Comparison of the flood damage cost estimate of the normal control strategy with a flood risk averse AWM strategy 119

Figure 4.28 Comparison of the total damage estimate of the normal control strategy with a flood risk averse AWM strategy 120

Figure 4.29 Theoretical potential of total cost reduction by applying AWM with perfect (synthetic) forecasts to the Rijnland water system for extreme events with return periods between 10 and 100 years. Note the logarithmic scale of the cost-axis. 121

Figure 4.30 Optimisation of control horizon by minimising estimated total damage costs. Both the flood costs and the total costs become stable after 70 hrs. Expansion of the control horizon beyond 70 hrs has no use. The analysis has been performed on the basis of perfect (synthetic) forecasts for a 1/100 year event. 121

Figure 4.31 Estimatied drought (too low water levels) and flood (too high water levels) damage costs for AWM strategies generated with NSGAII optimisation. Costs are evaluated for the period between 1-9-1997 and 24-4-2004. The lower-left corner of the Pareto front shows strategies with total cost reductions of around 2.4*105 Euro compared to strategies without anticipation (Van Andel et al., 2009[b]). 123

Figure 4.32 Estimated Flood damage costs versus Probability threshold for the 150 least total cost AWM strategies determined by NSGAII optimisation. Costs are evaluated for the period between 1-9-1997 and 24-4-2004. The graph clearly shows how flood damage reduces when low probability thresholds are applied. Meaning that when only 1 or 2 ensemble forecast members are required to exceed the warning level, then most critical events will be identified and the forthcoming damage reduced by AWM strategy 123

Figure 5.1 Ribb and Gumara catchments with the locations of hydro-metrological gauging stations. 129

Figure 5.2 Data gaps of Addis Zemen gauging station, indicated by circles for long periods of missing data and an arrow for a short period of missing data. In the analysis only the data between 2000 and 2006 was used. 130

Figure 5.3 Gumara and Ribb daily river discharge from 1998 to 2006. The discharge data of the Ribb river in the wet season of 1998 seems too flat (circled), pointing to measurement errors. The data between 2000 and 2006 was used for analysis. 131

Figure 5.4 Gumara calibration result for 2001 133

Figure 5.5 Ribb calibration for 2001 133

Figure 5.6 Gumara validation for 2005 134

Figure 5.7 Ribb validation for 2005 134

Figure 5.8 Validation results: Ribb correlation (a) , Gumara correlation (b) 134

Figure 5.9 Flood areas around Lake Tana sub basin (Aug 13-27, 2006) (DFO, 2008) 136

Figure 5.10 River cross-sections of Ribb and Gumara 137

Figure 5.11 Gumara and Ribb river flows comparison 137

Figure 5.12 Example streamflow hindcasts with a 4-day forecast horizon. After 4 days the underestimation by assuming no rainfall becomes clear. Assuming Monthly mean rainfall shows a better comparison with the reference streamflow (simulated by using measured rainfall as input), but streamflow peaks, particularly in the beginning of the wet season, are underestimated. 139

Figure 5.13 NRSME for 1, 2, and 3-day forecast horizons for the year 2000 140

Figure 5.14. Gumara streamflow forecasts with Monthly mean precipitation forecasts as input to HEC-HMS, 3-day forecast horizon, wet season 2000 141

Figure 5.15. Gumara streamflow forecasts with Min EPS precipitation forecasts as input to HEC-HMS, 3-day forecast horizon, wet season 2000 141

Figure 5.16. Gumara streamflow forecasts with Mean EPS precipitation forecasts as input to HEC-HMS, 3-day forecast horizon, wet season 2000 142

Figure 5.17 Number of hits and false alarms with mean EPS 143

Figure 5.18 Mean EPS based flood forecast (2002, 2-days forecast horizon) 143

About the author

Schalk Jan van Andel was born 10th of May, 1978, in Amsterdam, The Netherlands. He lived in Kamoto, Zambia, for three years, and has worked as a lecturer at the secondary school of Fowakabra, Ghana. In 2003 he graduated (with distinction) for his MSc degree in Integrated and quantitative water management from Wageningen University. This programme included courses in hydrology, computational hydraulics and water management. During his MSc study he has been involved in national and international research projects, like the design of innovative flood reduction measures along the Dutch branches of the Rhine at Delft Hydraulics (now Deltares) and the development of Earth System Models at the Potsdam-Institut für Klimafolgenforschung (PIK). He specialised in the development and application of hydrological and hydrodynamic models.

After graduating he worked as a specialist water management with HydroLogic, The Netherlands, and as a project officer with the Netherlands Water Partnership (NWP). By the end of 2004 he joined UNESCO-IHE with the Hydroinformatics and Knowledge Management department and the Hydroinformatics core, to start the PhD research presented in this dissertation. He has published several papers in meteorological and hydrological international journals, and is a member of the international HEPEX initiative on Hydrological Ensemble Prediction EXperiments.

At present Schalk Jan is a lecturer in Hydroinformatics at UNESCO-IHE, Delft, The Netherlands. He is involved in a number of national and international research projects on operational water management and real-time control of water systems. His research interest concerns the application of meteorological data and forecasts in operational water management.

Publications in peer-reviewed journals

Andel, S.J. van, Price, R.K., Lobbrecht, A.H., Kruiningen, F. van, Mureau, R., 2008[a]: Ensemble Precipitation and Water-Level Forecasts for Anticipatory Water-System Control, J. Hydrometeor., 9, 776-788.

Andel, S.J. van, Lobbrecht, A.H., Price, R.K., 2008[b]: Rijnland case study: hindcast experiment for anticipatory water-system control, Atmospheric Science Letters, Vol. 9, No 2, 57-60

Andel, S.J. van, Price, R.K., Lobbrecht, A.H., Kruiningen, F. van, 2009[a]: Modelling controlled water systems, J. of Irrigation and Drainage, *in press*

Andel, S.J. van, Price, R.K., Lobbrecht, A.H., Kruiningen, F. van, Mureau, R., 2009[b]: Framework for Anticipatory Water Management: application in flood control for Rijnland reservoir system, *submitted*

Akhtar, M.K., Corzo, G.A., Andel, S.J. van, Jonoski, A.: River flow forecasting with artificial neural networks using satellite observed precipitation pre-processed with flow length and travel time information: case study of the Ganges river basin, Hydrol. Earth Syst. Sci., 13, 1607-1618, 2009

Conference papers

Lobbrecht, A.H., Andel, S.J van, 2005: Integrated urban and rural water management using modern meteorological data, Proc. 10th International Conference on Urban Drainage, 21-26 August 2005, Copenhagen, Denmark

Andel, S.J. van, Lobbrecht, A.H., 2005: Ensemble weather forecasts - Applicability and use in flood prevention, Proc. Actif conference, Innovation, advances and implementation of flood forecasting technology, 17 to 19 October 2005, Tromsø, Norway

Andel, S.J. van, Lobbrecht, A.H., 2006: Ensemble weather forecasts and operational management of regional water systems, 7th International Conference on Hydroinformatics (ed. by P. Gourbesville, J. Cunge, V. Guinot & S.Y. Liong), Research Publishing Services, 1351-1358, Nice, France

Lobbrecht, A.H., Andel, S.J. van, Kruiningen, F. van, 2006: Operational management of hydrological extremes using global-scale atmospheric models, in: Climate Variability and Change—Hydrological Impacts (Proceedings of the Fifth FRIEND World Conference held at Havana, Cuba, November 2006), IAHS Publ. 308, 2006., Havana, Cuba

Griensven, A. van, Akhtar, M.K., Haguma, D., Sintayehu, R., Schuol, J., Abbaspour K., Andel, S.J. van, Price, R.K., 2007: Catchment Modelling with Internet based Global Data, 4th International SWAT conference, July 2-7, Delft, the Netherlands

Andel, S.J. van, Lobbrecht, A.H., Price, R.K., 2007: Rijnland case study: anticipatory control of a low-lying regional water system, in Thielen., J., J. Bartholmes J., and J. Schaake (Eds.) (2007) 3rd HEPEX workshop, Book of Abstracts, European Commission EUR22861EN

Andel, S.J. van, Lobbrecht, A.H., Price, R.K., 2008: Anticipatory Water Management; cost-benefit analysis, Geophysical Research Abstracts, Vol. 10, EGU2008-A-06996, 2008

Andel, S.J. van, 2008: Anticipatory water management for advanced flood control, in Flood Risk Management: Research and Practice – Samuels et al. (eds), Taylor & Francis Group, London

S.Loos, S.J.van Andel, A.H.Lobbrecht, R.K.Price, 2008: Anticipatory water management, decision support for real-time operational and long term strategic use of new meteorological forecast products in flood control, Hydropredict, International Interdisciplinary Conference on Predictions for Hydrology, Ecology, and Water Resources Management: Using Data and Models to Benefit Society, Czech Republic

Assefa, K.A., Andel, S.J., Jonoski, A., Lobbrecht, A.H., 2009: Combining Different Verification Methods for Analysis of Flood Early Warnings: Fogera Plain, Lake Tana, Upper Blue Nile Case Study, 7th ISE & 8th HIC, Chile, 2009

Samenvatting

Water is nauw verweven met onze leefomgeving. Ontwikkelingen in onze maatschappij hebben via de ruimtelijke ordening invloed op het watersysteem. De watergerelateerde omgeving waarin we leven legt op zijn beurt beperkingen op aan het gebruik van de ruimte om ons heen. We richten onze leefomgeving zodanig in dat we onder normale omstandigheden goed gebruik kunnen maken van het water, zonder dat het water overlast veroorzaakt. Extreme omstandigheden kunnen echter tot problemen leiden met overstromingen en droogte als gevolg. Deze kritische gebeurtenissen kunnen worden geclassificeerd in te veel water, te weinig water, of water van een slechte kwaliteit. Met het waterbeheer trachten we voortdurend de frequentie en omvang van de schade die het gevolg is van kritische gebeurtenissen, te minimaliseren. We onderscheiden in het algemeen strategisch waterbeheer en operationeel waterbeheer. Strategisch waterbeheer is verweven met het landgebruik in een stroomgebied en de ruimtelijke ordening en hierbij spelen aspecten van het ontwerp van het watersysteem en effecten op lange termijn. Operationeel waterbeheer, het onderwerp van dit onderzoek, richt zich op de dagelijkse beheersing van het watersysteem, waarbij inzet van regelkunstwerken van belang is.

Een grote groep aan kritische gebeurtenissen in het waterbeheer zijn meteorologisch van aard. Het komt regelmatig voor dat waterbeheerders te laat zijn geïnformeerd over een op handen zijnde kritische gebeurtenis - zoals extreme neerslag - om daarop nog effectief te kunnen reageren. De tijd die beschikbaar is tussen het moment van een hydrologische of een meteorologische meting en het moment dat de waterbeheerder deze meting tot zijn beschikking heeft en kan ingrijpen, is te kort.

Weersverwachtingen en voorspellingen van het gedrag van het watersysteem bieden uitkomst en kunnen worden gebruikt om de beschikbare reactietijd voor de waterbeheerder te vergroten. De periode waarover vooruit kan worden gekeken noemen we de voorspellingshorizon. Het beheersen van het watersysteem, op basis van een hydrometeorologische voorspelling, wordt 'anticiperend waterbeheer' genoemd. Anticiperend waterbeheer stelt waterbeheerders in staat om op tijd maatregelen te nemen om de schade van kritische gebeurtenissen te beperken. Een voorbeeld van een anticiperende maatregel is het verlagen van de waterstand in een boezemstelsel om een overstroming te voorkomen, wat ook wel wordt aangeduid met 'voormalen'.

Zoals schakers die hun kansen op het winnen van het spel vergroten door op zetten van hun tegenstander te anticiperen, kunnen ook waterbeheerders de prestaties van hun watersysteem verbeteren door zich voor te bereiden op aankomende gebeurtenissen zoals extreme neerslag, hoogwater, overstromingen of juist droogte en slechte waterkwaliteit.

De hydrometeorologische voorspellingen, die voor anticiperend waterbeheer worden gebruikt, zijn niet altijd correct en zijn omgeven met een mate van onzekerheid. Die onzekerheid hangt samen met de weersverwachting en met de berekening van het effect van het verwachte weer op het watersysteem. Vooral de weersverwachtingen hebben een hoge mate van onzekerheid, omdat de atmosfeer, waarin het weer zich afspeelt, een chaotisch systeem is, waarin kleine verstoringen snel kunnen uitgroeien tot een niveau waarop ze ook grootschalige invloed hebben. Anticiperend ingrijpen met waterbeheerstechnische maatregelen is daardoor in de praktijk een complexe taak. Als gevolg van de onzekerheid in de voorspellingen en de complexiteit van waterbeheersing zullen maatregelen soms niet op tijd worden genomen, of achteraf niet nodig blijken te zijn geweest. Omdat anticiperende maatregelen mogelijk nadelige effecten met zich meebrengen, moeten de onzekerheid van de voorspelling en de risico's van een achteraf onjuist ingrijpen, worden meegenomen bij anticiperend waterbeheer.

De onzekerheid van de weersverwachting en de veranderlijkheid daarvan in de tijd, kan worden geschat met zogenaamde 'ensemble' voorspellingssystemen. Bij een ensemble voorspelling wordt een kansverwachting samengesteld, waarmee voor een zekere tijd vooruit de nauwkeurigheid van de voorspelling berekend is. Deze verwachting wordt bepaald door het computermodel dat wordt gebruikt voor de weersverwachting herhaaldelijk te draaien met variërende beginwaarden. Dit wordt op zo'n manier gedaan dat de variatie van de modeluitkomsten een maat is voor de onzekerheid van de verwachting. Op deze manier kunnen we rekening houden met het feit dat we maar in beperkte mate in staat zijn de actuele staat van de atmosfeer nauwkeurig te meten of te schatten.

Overheden en bedrijven in de watersector maken in toenemende mate gebruik van deze ensemble voorspellingen. De kansverdeling die voor iedere verwachting beschikbaar is, stelt een waterbeheerder in staat de risico's van mogelijke waterbeheerstechnische maatregelen mee te nemen in zijn beslissing om een maatregel al dan niet in te zetten. Veel onderzoek richt zich op het leveren van zo goed mogelijke hydrometeorologische ensemble voorspellingen. Het voorliggende onderzoek, richt zich juist op het zo effectief mogelijk gebruiken van beschikbare ensemble voorspellingen voor anticiperend waterbeheer.

In dit onderzoek is een raamwerk opgesteld voor het ontwikkelen van beheersstrategieën voor anticiperend waterbeheer. In de eerste plaats wordt in dit raamwerk nadruk gelegd op de beschikbaarheid van instrumenten uit de hydroinformatica, die het mogelijk maken om op flexibele wijze computersimulaties uit te voeren van gereguleerde watersystemen. Door gebruik te maken van dergelijke simulatiemodellen, kan een gangbare waterbeheerstrategie worden nagebootst en worden vergeleken met alternatieve, anticiperende strategieën.

In de tweede plaats wordt in het raamwerk benadrukt dat waterbeheerders zelf de prestaties van de hydrometeorologische voorspellingen voor hun beheersgebied zouden moeten verifiëren, waarbij achteraf een vergelijking wordt gemaakt met metingen. In het bijzonder geldt voor de meteorologische verwachtingen dat een verificatie voor het eigen beheersgebied nodig is omdat de prestatiescores, die worden geleverd door de meteorologische instituten, vaak zijn bepaald voor een regionale of wereldwijde schaal. Voor een lokaal stroomgebied kan de prestatie van de weersverwachtingen anders zijn.

Het is nodig om een verificatie op maat uit te voeren om de effectiviteit van anticiperend waterbeheer te kunnen vaststellen. Dit betekent bijvoorbeeld voor hoogwaterbeheersing dat de verificatie zich moet richten op neerslag en neerslag-afvoermodellering. Daarnaast moet de verificatie niet zijn gebaseerd op een vast tijdsinterval (bijvoorbeeld van een dag) maar op gebeurtenissen (bijvoorbeeld een extreme neerslaggebeurtenis die enkele dagen aanhoudt).

De verificatie moet worden uitgevoerd met continue meerjarige tijdreeksen en simulaties, en niet op basis van een aantal geïsoleerde kritische gebeurtenissen, zoals tot voor kort gebruikelijk was. Alleen met continue simulatie kunnen de volledige gevolgen van het toepassen van anticiperend waterbeheer worden bepaald, waaronder ook de risico's van onnodige alarmeringen. Dit laatste is een 'false alarm', of ook wel een waarschuwing voor een kritische gebeurtenis terwijl deze in werkelijkheid niet blijkt op te treden.

De beschreven aanpak duiden we ook wel aan met 'verificatie-analyse'. Hiervoor zijn historische meetreeksen van variabelen in het watersysteem, historische meteorologische gegevens en historische weersverwachtingen nodig. Als er geen historie van weersverwachtingen beschikbaar is, moeten die historische weersverwachtingen alsnog worden gegenereerd. Dit kan worden gedaan door het numerieke weersverwachtingsmodel opnieuw te draaien voor de analyseperiode. Dit wordt ook wel aangeduid met 're-forecasting' of 'hindcasting'.

Door de resultaten van de historische weersverwachtingen toe te passen op een simulatiemodel van het watersysteem, is het mogelijk om voor situaties uit het verleden alternatieve anticiperende waterbeheerstrategieën na te bootsen. Deze simulaties laten waterbeheerders zien wat er gebeurd zou zijn als ze in het verleden weersverwachtingen hadden gebruikt bij het operationele waterbeheer. Hiermee kan de effectiviteit van anticiperend waterbeheer voor kritische gebeurtenissen worden vastgesteld.

Voor veel overheden in de watersector zal zicht op een verbeterde effectiviteit van het waterbeheer alleen, niet voldoende zijn om te besluiten deze techniek toe te passen. In de meeste gevallen zal ook de efficiëntie moeten worden aangetoond. Omdat waterbeheer zeer dynamisch is, kunnen geen vaste kosten-baten verhoudingen worden gebruikt voor deze efficiëntie-analyse. Elke gebeurtenis is anders dan een vorige en hierdoor is ook de kosten-baten verhouding steeds weer anders.

Daarom wordt in het raamwerk voor anticiperend waterbeheer, in de derde plaats, benadrukt dat de waterbeheerder een kostenmodel zou moeten opstellen voor de relatie tussen toestandsvariabelen - zoals waterstanden - en de opbrengsten of schade in het watersysteem. Hiermee kan de continue simulatie van de waterbeheersing worden vertaald in een tijdreeks van kosten. De totale kosten van kritische gebeurtenissen, en de ontwikkeling van deze kosten door de jaren heen, kunnen worden bepaald en worden vergeleken voor verschillende voorspellingsproducten en strategieën voor anticiperend waterbeheer.

Als uit het bovenstaande blijkt dat anticiperend waterbeheer efficiënt is, dan kan als extra analyse de beheerstrategie worden geoptimaliseerd. Een belangrijk doel van deze optimalisatie is het minimaliseren van de kosten van kritische gebeurtenissen, en tegelijkertijd het minimaliseren van de totale kosten over een meerjarige periode, waarbij ook de reguliere situaties en onjuist voorspelde situaties horen. Voor de optimalisatie van de operationele maatregelen lijkt het voor de hand te liggen gebruik te maken van een minimale risicobenadering, waarbij de kansverdeling uit de ensemble verwachting wordt gebruikt. Een belangrijke reden waarom deze techniek hier echter niet is gebruikt is dat de ensemble methode weliswaar een schatting van de kansverdeling geeft, maar dat dit niet altijd representatief hoeft te zijn voor de werkelijke kansverdeling. Dit probleem wordt ondervangen door van een meerjarige periode de effecten van het gebruik van ensembles te analyseren, waarbij alle onnauwkeurigheden in de verwachtingen impliciet zijn meegenomen. Het meerjarige optimalisatieprobleem, waarbij per dag meerdere ensemble kansverdelingen van toepassing zijn en de beste maatregelenstrategie voor de hele periode moet worden bepaald, is niet in een mathematisch optimalisatiemodel te vatten en daarom is gekozen voor een 'random search' methode, in dit geval een genetisch algoritme. Belangrijk is dat deze methode ook problemen aankan met meer doelfuncties tegelijk en daarmee een reeks alternatieve beheerstrategieën genereert. Zo wordt aan de waterbeheerders de mogelijkheid geboden om de door hen gewenste waterbeheerstrategie te selecteren, afhankelijk van de afweging van het belang van de onderkende doelfuncties. Een waterbeheerder kan er bijvoorbeeld voor kiezen een praktische strategie toe te passen waarbij op optimale wijze de verwachte kosten van hoogwatergebeurtenissen worden teruggebracht, terwijl de verwachte totale kosten daardoor misschien niet op het minimum liggen.

Het raamwerk voor het ontwikkelen van strategieën voor anticiperend waterbeheer is toegepast op twee praktijkonderzoeken die betrekking hebben op hoogwatervoorspelling, -alarmering en -beheer. Eén van deze onderzoeken betrof het polder-boezemsysteem van het Hoogheemraadschap van Rijnland in Nederland. De ander betrof een deelstroomgebied van de Blauwe Nijl, bovenstrooms van Lake Tana in Ethiopië. De ensemble neerslagvoorspellingen van het ECMWF Ensemble Prediction System (EPS) en het NCEP Global Forecasting System (GFS) zijn in de onderzoeken gebruikt. Het EPS wordt al operationeel ontvangen door het hoogheemraadschap. Het GFS is vrij beschikbaar via het internet, wat het een zeer interessant onderzoeks- en operationeel instrument maakt voor ontwikkelingslanden, met beperkte investeringscapaciteit in meteorologische verwachtingsinformatie.

In het Nederlandse praktijkonderzoek werden de meeste hoogwatergebeurtenissen in de meerjarige analyseperiode van 8 jaar goed voorspeld. De optimalisatie van de anticiperende waterbeheerstrategie voor Rijnland resulteerde in een reductie van 35% van de hoogwaterschade en een reductie van 30% in de totale schade. Dit laat zien dat anticiperend waterbeheer tot beduidend betere resultaten leidt dan het traditionele reactieve operationele waterbeheer. In Nederland kunnen de ECMWF EPS voorspellingen worden gebruikt om de voorspellingshorizon uit te breiden tot drie dagen of meer. De aanmerkelijke verschillen tussen de gevonden optimale beslissingsregels en de beslissingsregels die op het moment door Rijnland worden toegepast, bevestigen dat het toepassen van hindcasting-analyses positieve resultaten oplevert voor het verbeteren van anticiperende waterbeheerstrategieën.

De resultaten van het praktijkonderzoek in de Blauwe Nijl laten zien dat vrij beschikbare weersverwachtingen en hydrologische modellerings-software goed kunnen worden gebruikt bij onderzoek naar voorspellingssystemen en strategieen voor anticiperend waterbeheer. In een gebied waar momenteel geen vorm van waarschuwing voor hoogwater beschikbaar is, is dat van grote waarde. In dit specifieke praktijkonderzoek kon maximaal 60% van de gesimuleerde referentie hoogwatergebeurtenissen worden voorspeld. Het voorspellingssysteem moet eerst verder worden verbeterd voordat operationeel gebruik realistisch is. Bij het realiseren van deze verbeteringen moeten bias-correctie en neerschalingsmethoden worden gebruikt. Een brede internationale onderzoeksgemeenschap op het gebied van ensemble voorspellingen richt zich op de ontwikkeling van deze methoden. Deze methoden, die tot doel hebben om betrouwbare probabilistische voorspellingen te genereren met een zo klein mogelijke onzekerheidsmarge, worden thans ook gebruikt om de efficiëntie van anticiperend waterbeheer voor waterschappen in Nederland nog verder te vergroten.

Het belangrijkste onderdeel bij het ontwikkelen van succesvolle anticiperende waterbeheerstrategieën, is de verificatie-analyse met continue simulaties voor perioden van verscheidene jaren. Consistente meerjarige historische reeksen met weersverwachtingen zijn noodzakelijk vanwege de lage frequentie van kritische gebeurtenissen. Dergelijke historische gegevens zijn over het algemeen niet beschikbaar omdat weersverwachtingsystemen continu worden vernieuwd en er tot voor kort binnen de meteorologie weinig aandacht was voor systematische opslag van deze informatie. Er is daarom een grote behoefte aan het creëren van hindcast reeksen ter ondersteuning van de ontwikkeling van nieuwe toepassingen, zoals in het waterbeheer. Omdat het generen van hindcasts conflicteert met de operationele taken van meteorologische instituten, en wel in verband met de beschikbare mens- en computercapaciteit, moet de taak van het hindcasten liefst aan aparte, gespecialiseerde organisaties worden overgelaten die hier onafhankelijke financiering en rekencapaciteit voor beschikbaar hebben. Dit zal een grote bijdrage leveren aan het praktisch gebruik van weersverwachtingen voor operationeel waterbeheer.

Het gebruik van weersverwachtingen is belangrijk voor het waterbeheer wereldwijd. Deze verwachtingen zijn tegenwoordig van zodanige kwaliteit, dat de vraag gesteld moet worden of waterbeheerders het zich nog kunnen veroorloven geen gebruik te maken van deze beschikbare informatie. Dat geldt niet alleen voor het hoogwaterpraktijkonderzoek dat in dit proefschrift is gepresenteerd, maar voor veel meer toepassingen, zoals droogtebeheer, irrigatie, en voor een breed scala aan typen watersystemen. Daarom wordt hier een beroep gedaan op wetenschappers, ingenieurs en waterbeheerders om gezamenlijk het volledige toepassingsbereik van anticiperend waterbeheer te ontwikkelen en het gebruik van hydrometeorologische voorspellingen in het operationele waterbeheer te stimuleren.